Springer Theses

Recognizing Outstanding Ph.D. Research

Author
Dr. Bo Jing
Hefei National Laboratory for Physical Sciences at the Microscale
University of Science and Technology of China
Hefei, Anhui, China

Supervisor
Prof. Xiao-Hui Bao
Hefei National Laboratory for Physical Sciences at the Microscale
University of Science and Technology of China
Hefei, China

ISSN 2190-5053 ISSN 2190-5061 (electronic)
Springer Theses
ISBN 978-981-19-0327-4 ISBN 978-981-19-0328-1 (eBook)
https://doi.org/10.1007/978-981-19-0328-1

This Springer imprint is published by the registered company Springer Nature Singapore Pte Ltd.
The registered company address is: 152 Beach Road, #21-01/04 Gateway East, Singapore 189721, Singapore

Supervisor's Foreword

Quantum network has attracted much attention due to its potential applications in quantum communication, quantum computation, quantum metrology, etc. A starting point in this direction is a multiqubit entanglement shared among several remote quantum memories, which requires the incorporation of a number of demanding techniques such as local entanglement generation, efficient light-matter interface, and high-fidelity photon interference.

In pursuing such a goal, this thesis focuses on the system of cold atomic ensembles and performs in-depth experimental investigations. The author firstly gives a brief review on state of the art and presents experimental details on how to entangle a single photon with an atomic ensemble, which are very useful for someone new to this research field. Afterwards, the author presents several key achievements in improving the single-node performance, such as improving the intrinsic retrieval efficiency to 0.9, realizing filter-free operation by setting the different resonant conditions of photon-cavity, and achieving an atom-photon entanglement source with high overall efficiency. Furthermore, by using three similar setups, the author reports the first realization of GHZ entanglement of three separate quantum memories, a building block for 2D quantum repeaters, which is a milestone for quantum networks. Finally, by combining quantum memories and a time-resolved measurement, the author reports the first multi-photon interference experiment with different colors, which may find applications in the connection of dissimilar quantum nodes made of different physical systems.

The research described in this thesis makes a key contribution to the atomic ensemble-based approach of quantum networks. It may become a valuable resource for a broad readership to study efficient quantum memory, atom-photon entanglement generation, quantum networks, etc.

Hefei, Anhui, China | Prof. Xiao-Hui Bao
December 2021

Abstract

Realization of quantum entanglement among multiple remote quantum memories lays the foundations of building quantum networks, which enable multi-party quantum communication, distributed quantum computing, distributed quantum precision measurement, fundamental tests of physics under large-scale space, etc.

As photon can transfer quantum information for long distance, it suits well for connecting different quantum memories. Generating entanglement between photon and quantum memory with high quality would be important. There are many physical systems suiting for quantum memory, while quantum memory with cold atomic ensembles is favored for its superiority in terms of retrieval efficiency and storage lifetime. In cold atomic ensembles, photon-atom entanglement can be generated directly, which suits very well for quantum networks. In this paper, we mainly focused on the research of quantum memory with cold atomic ensembles. By using the ring cavity, the interaction between atoms and photons can be enhanced, which helps increase the intrinsic retrieval efficiency, and the filter device can be also excluded by setting the control field off-resonance from cavity while signal field in-resonance with cavity. Thus, we can acquire photon-atom entanglement with high brightness. After optimizing the influence of polarization, multi-excitation events, etc., we acquire atom-photon entanglement with high fidelity. Combining with the interference of identical photons, entanglement swapping can be realized with high fidelity. Three quantum network nodes are connected, GHZ-type entanglement of three cold atomic ensembles and hybrid entanglement of three cold atomic ensembles and three photons are generated for the first time. Furthermore, to extend the distance among quantum network nodes, quantum frequency conversion is used and non-classical correlation between photons at telecom band with atoms is observed. Then heralded entanglement between cold atomic ensembles separated by 50 km fiber is also created. Besides, by making use of three cold atomic ensembles, we generate three photons with different colors simultaneously and experimentally demonstrate time-resolved interference among the three photons, which paves the way to improving the compatibility of quantum networks in the future.

Our work of entanglement of multiple cold atomic ensembles and interference of frequency-distinguished photons would greatly promote the construction of quantum network. In the future work, deterministic atom-photon entanglement from the Rydberg system and long-lived quantum memory can be used to increase the success probability of nodes connection, which would promote the realization of a global quantum network.

Preface

Quantum information technology has been proven to have unique advantages over many fields and can help solve many problems that can't be processed with classical information technologies. To make it more practical, it is necessary to construct global quantum networks. Towards the realization of global quantum networks, my Ph.D. research is focusing on achieving the goal via entangling multiple quantum network nodes. In particular, I aim to build high-performance quantum memories with cold atomic ensembles and use them to perform explorations on multiple quantum memories.

In this thesis, I review the current state-of-the-art developments of quantum networks, detail abundant experimental skills of quantum optics, cold atoms, cavity-enhanced photon-atom interaction, etc., and report several excellent works, including filter-free quantum memory, multiple entanglement of quantum memories and single photons, interference of multiple frequency-distinguished photons from cold atomic quantum memories, etc. My works would be a big step towards quantum network-based multi-party quantum communication, distributed quantum computing, distributed quantum precision measurement, fundamental tests of physics under large-scale space, etc.

The thesis is organized as follows:

1. State of the art of quantum networks with various quantum memories.
2. Brief theories on the interaction between single photons and atomic ensembles with free space model or cavity model.
3. Experimental details on preparing cold atomic ensembles.
4. Experimental realization of efficient quantum memories with cold atomic ensembles with the help of ring cavity enhancement.
5. The first GHZ-type demonstration of entangling three separate quantum memories and entangling three separate quantum memories and three flying photons.
6. Experimental time-resolved interference of three photons with different colors.
7. Entanglement of two quantum memories via 50 km photon transmission.
8. Summary of my Ph.D. research and outlook for further developments.

The main innovations of the thesis are as follows:

1. We for the first time demonstrate a filter-free quantum memory with cavity assistant. An intrinsic retrieval efficiency of 0.9 and overall measured retrieval efficiency of 0.4 have been reported.
2. We report the first demonstration of GHZ entanglement among three separate quantum memories and hybrid entanglement of three quantum memories and three flying photons. The works pave a big step towards quantum network with multiple cold atomic quantum memories.
3. We perform time-resolved interference of three frequency-distinguished photons from three quantum memories, which opens a way for the connection of quantum memories with various physical platforms.
4. We demonstrate the longest fiber transmission distance for write-out photons via which two quantum memories are still entangled. The work sets a building block for long-distance quantum repeater.

Throughout the thesis, I try to present our experiments step by step with enough details, hoping the thesis be easier to understand if the readers follow the works from the beginning. The thesis can have a broad readership and benefit all students, researchers, and technical personnel who work in quantum information sciences.

Chengdu, Sichuan, China
December 2021

Dr. Bo Jing

Acknowledgements

I have spent nearly ten years at the University of Science and Technology of China for my studies, and now I am about to graduate with a Ph.D. A lot of deep emotions come into my mind right now. So many persons and things I met at the university are very important to my growth. I would like to express my sincere thanks while I am preparing the doctoral dissertation.

First of all, I would like to express my great gratitude to Prof. Xiao-Hui Bao, my doctoral supervisor. It would never be possible for me to complete the work without Prof. Bao's guidance and support. Professor Bao is very knowledgeable both in his understanding of quantum theories and his mastery of experimental skills. I am very impressed by his rich theoretical knowledge in research that he can always explain complicated formulas and theories with simple physical principles. In experiments, he can always inspire everyone to think divergently and put forward very simple but effective experimental methods. Professor Bao has not only achieved fruitful scientific research results in the key research fields of cold atom quantum memory and quantum optics, but also made achievements in circuit design, time sequences control system, data acquisition and recording, etc. Besides, Prof. Bao is also kind to his students. He is not only a teacher, but also like an elder brother to us. When we encounter difficulties in the experiments, he always encourages us to never give up and leads us to find a way to keep the experiments going forward. His optimistic attitude and perseverance in scientific research continue to inspire us to move forward. I am very grateful to have met such an excellent supervisor.

I would like to enthusiastically thank Prof. Jian-Wei Pan, the director of the Division of Quantum Physics and Quantum Information and one of the leaders in the field of quantum information in China. He has guided clear directions for each group of the laboratory and led us to continuously achieve innovative results.

Many thanks to Prof. Zhen-Sheng Yuan who led me into the field of cold atoms in my undergraduate stage. It was in his class that I started to take interest in cold atoms and learned a lot of theories about atoms and molecules.

Many thanks to Prof. Xiao Jiang, Dr. Na Li, and Dr. Xun Ding for providing me help in the fabrication and use of electronic devices needed in the experiments.

I would like to thank Prof. Jun Zhang and his students Dr. Wen-Hao Jiang and Mr. Yu-Qiang Fang. They improved the detection efficiency of the single-photon detector for us so that the brightness of photon detection in the experiments was enhanced and a lot of experiment time was saved.

Thanks to Dr. Yan Jiang and Dr. Jun Li for providing me guidance in building the experimental platforms.

I would like to thank Dr. Xu-Jie Wang, Mr. Yong Yu, and Mr. Peng-Fei Sun. It would never be easy to have good results without a nice teamwork with them.

I am grateful to Mr. Xi-Yu Luo, Mr. Chao-Wei Yang, Mr. Zi-Ye An, Mr. Jian-Long Liu, Miss Ren-Zhou Fang, and Miss Xiao-Yi Lai for their help in the work. I enjoyed the time studying with them in the Quantum Memory Research Group.

I would like to thank all other group members including Prof. Shuai Chen, Prof. Kai Chen, Prof. You-Jin Deng, Prof. Nai-Le Liu, Prof. Xi-Lin Wang, Dr. Sheng-Jun Yang, Dr. Tian-Ming Zhao, Mr. Guo-Peng Lu, Miss Ming-Ti Zhou, etc. I want to thank them all for their help and support in the experiments.

I also want to thank Mrs. Jin-Xiu Xia, Mrs. Chuan-Fang Wang, and Mrs. Hong-Hua Zhao for their help in purchasing experimental equipment and daily reimbursement.

I am grateful to Mr. Hai-Jun Pan and Mr. Guan-Yin Gao who gave me great help in my undergraduate and postgraduate stages.

I would like to thank the Virya Foundation for providing me help during the undergraduate stage. Particularly, I want to thank Prof. Zong-Guang Pan, the Chairman of the Foundation, who impressed me with his wisdom and encouraged me to face difficulties in life bravely and solve them dispassionately.

Finally, I want to express my gratitude to my family, including my beloved wife Mrs. Yun-Xia Zhao, my parents and sisters for their long-term company and support. They are always with me in the difficult moments of my life. I love them!

November 2018 Dr. Bo Jing

Contents

Chapter 1
Introduction

Since Planck's 'quantum' hypothesis was first proposed in the early 20th century, quantum mechanics has developed rapidly for more than a century, making people have a new understanding of the micro quantum world, and it has been applied to chemistry, biology, engineering electronics and other fields. Many new disciplines based on quantum science have also emerged from the developments. Over the past few decades, the combination of quantum mechanics and information sciences has induced a new research field: quantum information science [1–3]. Several countries have proposed detailed roadmap [4] to develop quantum science and technology, especially on quantum communication, quantum computing, quantum precision measurement, and so on [5–12].

Thanks to the no-cloning theorem [13] of quantum information, quantum communication has great advantages in confidentiality and security in contrast to classical communication. The realization of quantum communication among different countries and even the world is attracting the continuous efforts of scientists, one of the most important points is the ability to build a quantum network all over the world [14–17]. The construction of quantum networks will make many promising applications possible and promotes the research and development of related fields, including the research of quantum repeaters to achieve long-distance multi-party quantum communication [9, 15], distributed quantum computing [18], atomic clocks and precision measurement [19, 20], telescopes with improved resolution [21] and fundamental tests of quantum mechanics [22] in large-scale space.

The construction of a global quantum network means that there exists quantum correlation between any two nodes of the world and quantum information can be transferred between the nodes. An efficient way for such goals is to utilize quantum teleportation technology with entangled states for the transmission of quantum information [23, 24], thus the most important thing for the construction of quantum network is to realize quantum entanglement of separate network nodes. In this thesis, I will mainly focus on how to connect multiple quantum network nodes and construct quantum correlations of different nodes.

B. Jing, *Quantum Network with Multiple Cold Atomic Ensembles*, Springer Theses,
https://doi.org/10.1007/978-981-19-0328-1_1

Generally, to realize the connection between quantum network nodes with a distance of less than 500 km, it is possible to directly transmit photonic quantum information through optical fibers [15, 25]. However, for the distance of 500–2000 km, due to the loss of photons during transmission and the limitation of the quantum no-cloning theorem, direct transmission is no longer realistic. Under this situation, it is often necessary to rely on quantum repeaters [9, 26] to connect multiple short-distance nodes and entanglement swapping is used to realize the connection of distant nodes. For quantum nodes with a distance of more than 2000 km, the quantum repeater on the earth surface requires quantum memories with higher efficiency and fidelity when connecting short-distance nodes, which will also make the resource consumption greater. To let the quantum repeater scheme still work, it is often necessary to utilize multiple satellites which serve as quantum network nodes to realize quantum repeaters both in outer space and earth. For realization of practical quantum repeater, one of the most key technologies is to build a high-performance quantum memory [27–30]. Under the help of quantum memories, the entanglement establishment and entanglement swapping between multiple short-distance nodes in the quantum repeater no more need to be successful at the same time. The successful entanglement of quantum nodes can be stored and wait for further connection after the remaining nodes have successfully established entanglement. After the entanglement connection between distant nodes is realized, quantum teleportation can be performed to realize quantum communication.

1.1 Quantum Teleportation

As shown in Fig. 1.1, direct transmission of a quantum state (encoded as $|\psi\rangle_A = \alpha|0\rangle_A + \beta|1\rangle_A$) at node A to a distant node C will suffer a great photon transmission loss, which will significantly affect the state transfer efficiency. However, assuming an entangled state $|\Psi^+\rangle_{BC}$ between node B (located in the same place as node A) and node C has been prepared earlier, i.e., quantum states of the three nodes can be written as:

$$\begin{aligned} |\psi\rangle_A \otimes |\Psi^+\rangle_{BC} = \frac{1}{2}[&|\Phi^+\rangle_{AB}(\beta|0\rangle_C + \alpha|1\rangle_C) \\ &+ |\Phi^-\rangle_{AB}(-\beta|0\rangle_C + \alpha|1\rangle_C) \\ &+ |\Psi^+\rangle_{AB}(\alpha|0\rangle_C + \beta|1\rangle_C) \\ &+ |\Psi^-\rangle_{AB}(\alpha|0\rangle_C - \beta|1\rangle_C)] \end{aligned} \tag{1.1}$$

In the equation,

$$\begin{aligned} |\Psi^\pm\rangle &= \frac{1}{\sqrt{2}}(|0\rangle|1\rangle \pm |1\rangle|0\rangle) \\ |\Phi^\pm\rangle &= \frac{1}{\sqrt{2}}(|0\rangle|0\rangle \pm |1\rangle|1\rangle) \end{aligned} \tag{1.2}$$

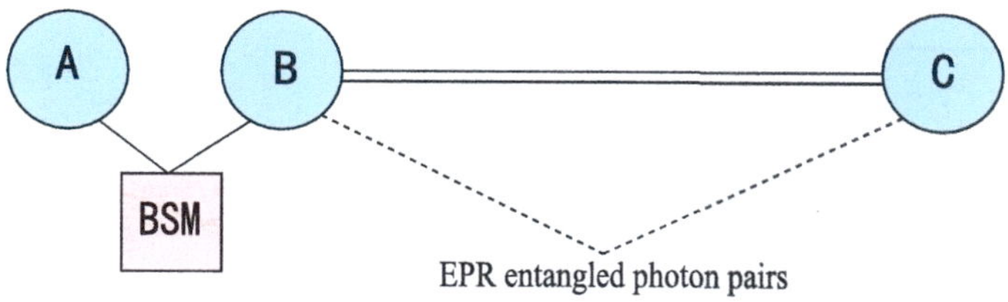

Fig. 1.1 Schematic for quantum teleportation

represent the four Bell states, which constitute a complete basis in a two-dimensional Hilbert space. By performing Bell state measurements on qubits at node A and B, quantum states between A and B will collapse to one of the four Bell states. The remaining state at C will collapse to a related state described in Eq. 1.1 and carries the qubit information α, β. To recover the qubit information as initial state prepared at A, transmission of the BSM results from A/B to C via classical communication channel is necessary and corresponding unitary operations on qubit at C will help recover the transferred qubit. In simple words, quantum information at node A can be securely transferred to node C via BSM on qubits at A and B and operation on qubit at C, which is the so-called quantum teleportation. As there exists transmission of classical information, the information isn't transferred exceeding speed of light, the whole process doesn't violate the theory of relativity. Since 1997 when Anton Zeilinger's group firstly reported experimental realization of quantum teleportation [31], there have several groups in the world announcing the experiments on quantum teleportation at different physical systems [32–38].

Quantum teleportation has been considered as an efficient way for realizing long distance quantum communication, the basic building block for quantum teleportation is the prepared entanglement of distant quantum network nodes. In another word, it is an significant step to entangle two or even more distant quantum network nodes, which is our main research interest in this thesis.

1.2 Connection of Two Quantum Network Nodes

The building block for quantum networks is to realize quantum connection between two network nodes and transmit quantum information. Up to date, there has been research group reported quantum key distribution around 400 km with fiber transmission [25], but extending of distance stays difficult, mainly limited by the photonic loss over long distance transmission. It still seems hard to realize quantum communication over thousands kilometers fiber transmission.

Closer to finally realize quantum communication, it is necessary to utilize photons with low transmission loss as flying qubits. A good choice is photons around 1550 nm (telecom C-band), which can undergo transmission loss as low as 0.2 dB/km with fibers, as shown in Fig. 1.2a. Specially, as an alkali metal, rubidium has a relatively

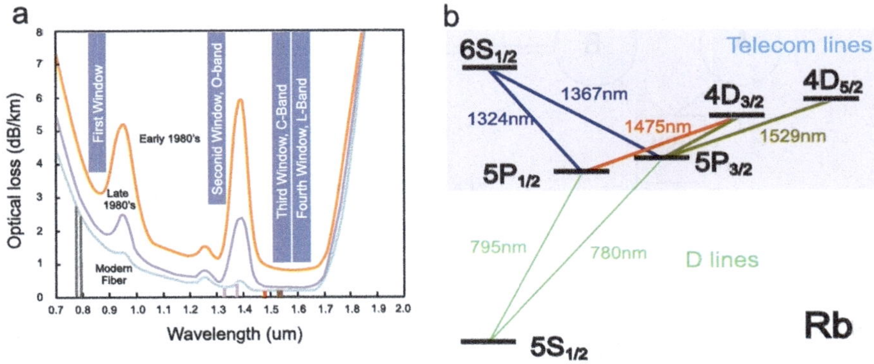

Fig. 1.2 **a** Optical loss over fiber transmission with different wavelengths. **b** Energy level structure of rubidium-87 atom and its corresponding wavelengths. Figures are reproduced from Ref. [43]

simple energy level, and the corresponding energy level structure can work for telecom band photons (Fig. 1.2b). Therefore, many experimental groups including us are based on rubidium to carry out the experiments. However, even with photons at telecom C-band, the transmitted signals will attenuate to 10^{-20} after transmission over 1000 km fibers, hardly to have any applications in real world. One solution to this type of problem is to use the small thickness of the atmosphere vertical to the sea level, which is about 8 km. Outside is nearly vacuum, and the photons in the vacuum are almost lossless. The main disadvantages are only when the distance becomes longer, the photon spot will diverge and become larger and larger, which increases the difficulties on photon collection and detection. Therefore, it is feasible to use multiple satellites outside the atmosphere to realize quantum communication [39]. It effectively reduces the optical loss that can't be avoided in the atmosphere or optical fiber. Nowadays, there are research groups which have reported quantum teleportation between satellite and ground with the physical separation distance exceeding 1200 km [40]. Other experiments such as entanglement distribution, quantum key distribution between satellite and ground are also demonstrated [41, 42]. An alternative way to overcome difficulties in direct photon transmission is analog to the way in classic communication, where repeater stations are used. In short words, after the photonic signal transmitted for a distance with low loss, copy and amplify the signal, then resend it and repeat such processes continuously through the repeater stations. Thus, the long-distance transmission of signals can be realized. Unfortunately, in quantum communication, the quantum no-cloning theorem limits the copying and re-transmission of signals in principle.

Inspired by the classical repeater and to apply the repeater scheme into quantum region, in 1998, H. J. Briegel, W. Dür, J. I. Cirac and P. Zoller proposed quantum repeater scheme [26] to realize long distance quantum communication, the basic idea is shown in Fig. 1.3. Since the photon transmission loss increases exponentially with the increase of distance, the transmission distance is limited to several hundreds of kilometers. To realize the connection between the quantum network nodes A and Z separated by a long distance L, the best way is to set up a repeater station, then the transmission distance of photons limited to the short distance between repeater

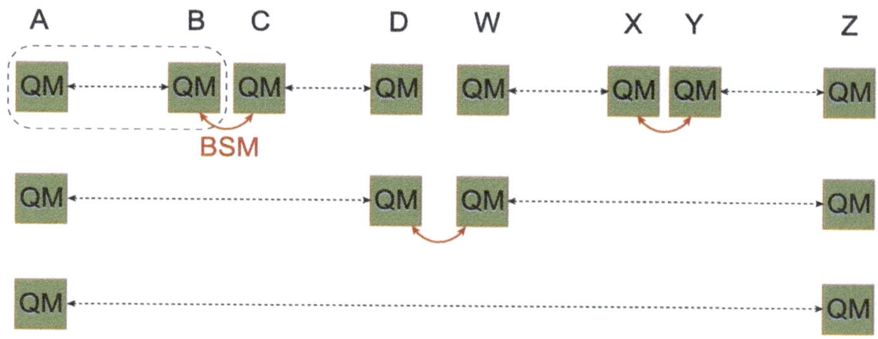

Fig. 1.3 Schematic for one-dimensional quantum repeater

stations is L_0 (such as distance between A and B). As shown in the figure, we can set up repeater nodes between A and Z and directly establish entanglement between A and B, C and D, W and X, Y and Z, then, then perform Bell state measurements (BSM) on nodes B, C and X, Y to achieve entanglement swapping to obtain the entanglement between A and D, W and Z, and finally perform BSM on D, W, as a result, entanglement between A and Z can be acquired, which can be further used to realize the transfer of quantum information from A to Z. During the whole process, the photon will no longer undergo the loss caused by direct transmission with distance L, but the loss with each photon's transmission distance L_0. The overall signal attenuation will decrease from scaling as exponential to polynomial, which is very beneficial for realizing connection of quantum network nodes with long distance.

1.3 Connection of Multiple Quantum Network Nodes

In the previous section, I introduced the use of a quantum repeater for realization of the connection between two distant quantum network nodes. However, such a linear quantum repeater can be only used for point-to-point communication. For real-world quantum networks, what is needed is not only quantum correlations between two liner repeater nodes, but also connections among multiple quantum network nodes. In Ref. [44], A Pirker et al. proposed schemes that can be used to connect multiple network nodes. One of the relatively simple schemes is to operate on multiple GHZ-type (Greenberger-Horne-Zeilinger) entanglement of three short-distance nodes and generate GHZ-type entanglement among three nodes with longer distance, which is the so-called 2D quantum repeater [45]. Further 2D quantum networks can be built based on these distant entanglements of three separate quantum networks nodes.

As shown in Fig. 1.4a, a simple schematic for 2D quantum repeater is presented. Through local operations on multiple GHZ states, three quantum networks nodes with longer distance can be entangled. Figure 1.4b shows the construction of quantum networks with GHZ states [44]. Small scale quantum networks are first built through operations on multiple GHZ states, larger scale quantum networks are built through

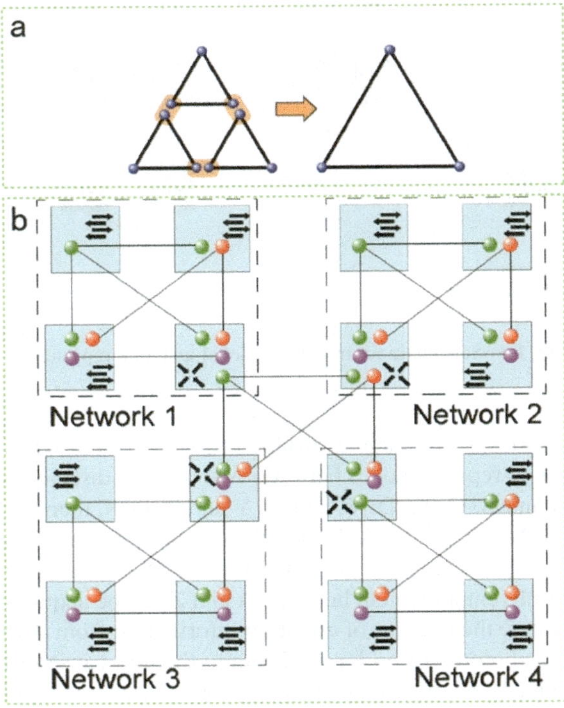

Fig. 1.4 Quantum networking with GHZ states. **a** Proposal of 2D quantum repeater. Entanglement of three distant quantum nodes are acquired from three short-distance GHZ states. **b** Schematic diagram of 2D quantum networks with GHZ states. Figure is reproduced from Ref. [44]

operations on neighboring nodes from different small scale quantum networks. To achieve the goal of building large scale quantum networks, the fundamental building block is to successfully generate entanglement of three quantum networks nodes, which is one of the main goals reported in this thesis.

During the process of using quantum repeaters to realize quantum connection of distant quantum networks nodes, if all the processes are carried out at the same time, the failure of any steps in the process will result in the whole program restarting from beginning, which is obviously not an efficient way. As a solution, it is best to perform the entanglement swapping at the next level or higher level after the entanglement of the previous level has been successfully established. In this way, even entanglement establishment of two neighboring nodes fails, we can only repeat the preparation process until entanglement between these two nodes are successfully established, and don't need to start all processes over. To make this proposal work, one important thing is to have the ability to store quantum states, and then wait to perform further entanglement swapping until the successful information of entanglement generation of previous levels is obtained. Therefore, quantum memory is of great significance in quantum networks, and it is particularly important to experimentally obtain the entanglement between photons and quantum memories with high retrieval efficiency, long storage time, etc.

1.4 Physical Realizations of Quantum Networks

Construction of quantum networks means realizing quantum entanglement of multiple quantum memories, so it is the first step to develop high-performance quantum memories. Quantum memories have a wide range of applications, in addition to being used in quantum networks, it also has great applications in other areas. In quantum computing, it is often necessary to operate qubits sequentially in time, or when multiple qubits need to be operated at the same time, time synchronization is required [6]. The requirement seems to hard to meet for all optical information processing, but quantum memories make this kind of operations very easy. It allows qubits to be stored in advance, and then be read out on time once the qubits need for further operation. In quantum precise measurement such as precise measurement based on atomic interferometer, the photonic quantum states can be first converted to atomic states and then read out, which can effectively attenuate the noise photons and improve the measurement accuracy of magnetic field, spectrum, etc. [46]. In experiments of generating multi-photon entanglement, it starts with single pair of entangled photons which are mostly prepared via probabilistic process (usually via spontaneous parameter down conversion , SPDC), the probability of that all pairs of photons exist at the same time and interfere for entanglement with more photons will decrease exponentially as the number of photons increases, it stays challenging for get entanglement with more and more photons. It was until last two years when ten-photon entanglement was reported [47, 48], in the ten-photon entanglement experiment, the overall photon coincidence counting rate is merely around several counts per hour, still too low to have substantial applications. However, under the help of quantum memories, multiple photons can be prepared at the same time via on-demand retrieval from quantum memories and applied quantum operations. In contrast to simultaneously preparing multiple photons through SPDC, the efficiency of multi-photons simultaneously prepared through quantum memories can be significantly improved [49, 50].

There have been many solutions for quantum memories, for instance, CRIB (Controlled Reversible Inhomogeneous Broadening)-based quantum memory [51, 52]. It mainly relies on the use of optical pumping technology to obtain a narrow absorption line, and then applying external controllable electric or magnetic field to let different atoms have different phase evolution, sometime later, flipping the electric or magnetic field to make atoms reverse the phase evolution and recover the initial atomic states, the absorbed photons will emit. In the CRIB scheme, if the external field distribution is controlled so that the resonance frequencies of the atoms are monotonously arranged in the crystal space position, a special CRIB scheme is obtained, namely GEM (Gradient Echo Memory) memory based on electromagnetic field gradient flip technology [53, 54]. CRIB-based quantum memory is severely limited by the external electric field and external magnetic field, and depends on the electromagnetic response characteristics of the storage medium, which is relatively troublesome compared to other quantum memory protocols. Another widely used protocol is AFC (Atomic Frequency Comb) quantum memory [55–57] based on

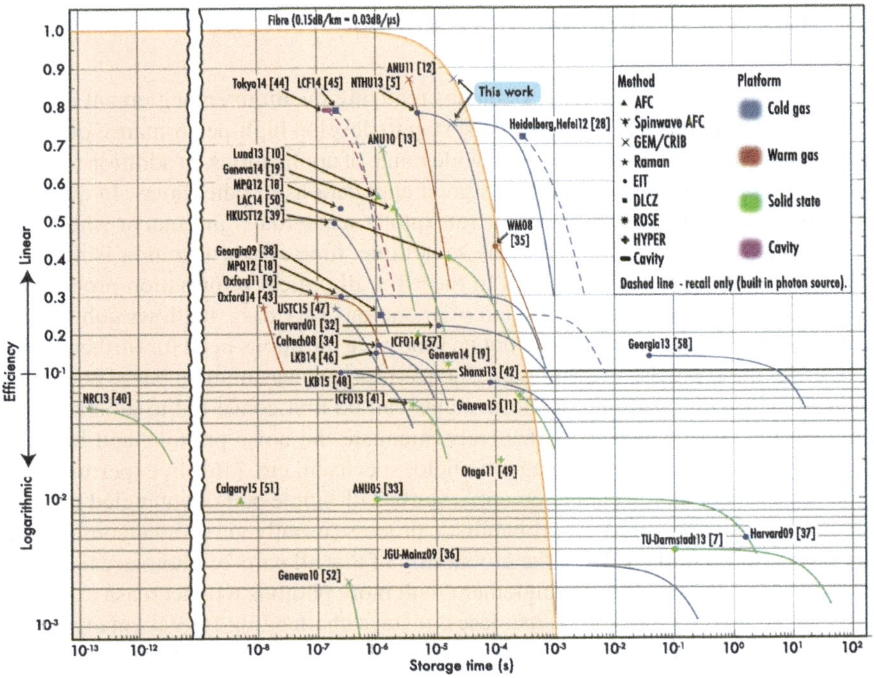

Fig. 1.5 Efficiency versus storage lifetime in quantum memories with different physical systems. Figure is reproduced from Ref. [65]

atomic frequency comb technology, which presents unique advantages in broadband and multiplexed quantum memory and suits very well for rare-earth ions doped solid state quantum memories.

For quantum memories with atomic gases, the main protocols include EIT (Electromagnetically induced transparency) quantum memory [58–61], Raman quantum memory [62, 63] and DLCZ quantum memory [64], etc. EIT quantum memory is used to store photons (signal photons) via coherently converting photonic states to atomic states, the detuning of signal photon from excited state is usually small, while the detuning is large for Raman quantum memory, which supports high-speed storage of photons. In this thesis, we mainly utilizes DLCZ-type quantum memory in cold atomic ensembles (CAEs), which will be introduced later in detail.

To assess the performance of a quantum memory for storage of quantum bits, several important indicators are used, including storage and retrieval efficiency, storage lifetime, storage fidelity, storage bandwidth, multi-mode storage capacity, etc. To acquire high-performance quantum memories, several groups are trying to improve the performance of quantum memories with different physical systems and realize the entanglement of remote quantum memories for quantum networking. For example, Rainer Blatt's group and Christopher Monroe's group explore quantum mem-

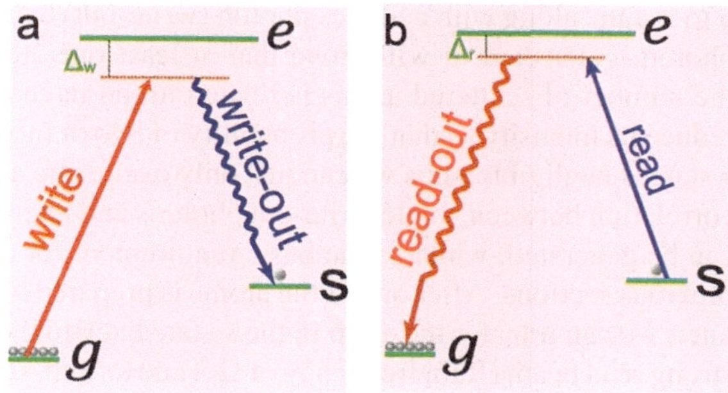

Fig. 1.6 Schematic for DLCZ quanutm memory. **a** Write process, **b** read process

ories with trapped ions [66–70], Gerhard Rempe's group and Harald Weinfurter's group explore quanutm memories with single neutral atoms [71–75], Ronald Hanson's group and Mikhail.D.Lukin's group explore quantum memories with NV/SiV centers [76–80], Nicolas Gisin's group, Matthew J.Sellars's group, Hugues de Riedmatten's group and Wolfgang Tittel's group explore quantum memories with rare earth ions doped solid state materials [81–84], Yoshihisa Yamamoto's group and Ataç Imamoğlu's group explore quantum memories with quantum dots [85–88], and some groups including our group focus on quantum memories with cold atomic ensembles [89–98]. In addition, there are also some other quantum memories based on warm atomic vapors [99–103], mechanical oscillators [104–106], etc. Different physical systems have their own advantages and disadvantages, some quantum memories have high efficiency but short storage lifetime, while some quantum memories may support long storage lifetime but have low memory efficiency.

As shown in Fig. 1.5, memory efficiency vs storage lifetime in different physical systems are compared. With current state of the art, quantum memories with cold atomic ensembles (right and top corner (Ref. [96]) in Fig. 1.5) are relatively at the forefront in terms of storage and retrieval efficiency, storage fidelity, and storage lifetime (Ref. [98] has reported lifetime up to sub-seconds). Under the use of cold atomic quantum memories, experiments reported in this thesis uses the entanglement between photons and cold atoms as a network interface to connect network nodes. The entanglement is generated through DLCZ protocol, so I will briefly introduce DLCZ protocol here.

As displayed in Fig. 1.6, for an atom with Λ energy structure, g, s are two ground states of atomic energy levels, e is the excited state, atoms are initially prepared in ground state g, the atomic states can be described as:

$$|\Psi\rangle_0 = |g \cdots g_j \cdots g\rangle. \tag{1.3}$$

At this time, a beam of weakly coherent light (write) is applied on the atomic ensemble, the detuning of write beam from the upper energy level e is Δ, the Rabi frequency is Ω_w. Through Raman scattering, the atoms will be probabilistically transferred from

g to s state, along with a Stokes photon (write-out) being emitted. Once the write-out photon is detected, it will prove that at least one atom has reached s state. Since the number of scattered atoms is related to the intensity of the write beam, we can reduce its intensity so that the probability of two or more atoms being scattered to the s state is negligible, thus we can just only excite one atom to s. As a result, quantum correlation between single write-out photons and internal states of atomic ensembles can be generated, which is the basic requirement for quantum repeater described in previous sections. After one of the atoms is prepared in the s state, some storage time later, we can transfer the atom in the s state back to the ground state g by applying a strong read beam (Rabi frequency of Ω_r) and an anti-stokes photon (read-out photon) will be released. By measuring the read-out photon, we can determine that the atom is indeed in the s state earlier, and by adjusting the storage time to read and detect, we can get the atomic coherence lifetime after being written to s state.

More dynamic process details of photon-atom interaction in the DLCZ quantum memory will be described in the Chap. 4.

1.5 State of the Art of Quantum Networks

To construct global quantum networks, the main idea is to realize entanglement of multiple long-distance quantum memories. However, up to date, no major breakthroughs have been made in realizing the entanglement of multiple quantum memories. The number of entangled quantum memories stays small. Most experiments mainly reported two-node entanglement and merely reported more entangled quantum memories. Entangled quantum memories where the number exceeds two are entanglement of four cold atomic quantum memories [107] demonstrate by K. S. Choi et al. in 2010 and entanglement of three atomic quantum memories [108] reported by Z.H. Yan et al. in 2017. In Ref. [107], as a further extension work of Ref. [109], the researchers split a single write beam into four beams and sent them into four cold atomic ensembles. Through DLCZ protocol that whether there is a write-out photon is correlated with whether the atomic states is excited, the researchers interfered the four write-out channels and erased the information which cold atomic ensembles the write-out photons comes from, and finally selected events that only one photon was detected. As there was no way to distinguish where the detected photons comes, i.e., which cold atomic ensembles was excited can't be distinguished, entanglement of the four cold atomic quantum memories was generated. This type of entanglement is W state where whether the atoms are excited and a spinwave is generated or not, the four atomic ensembles share a spinwave together, only one atomic ensemble is excited and has a spinwave, the other three atomic ensembles aren't excited and no spinwaves are generated. This kind of W state is sensitive to phase on the path, once the distance get longer, the phase stability requirements for the long distance will be very high, and the proportion of the vacuum state component (i.e., nothing happens to the quantum memories) is large, which will affect the fidelity of prepared entanglement when extending to more network nodes. The quality of the entangled state is far inferior to the maximum entangled state in application for global quan-

Table 1.1 Comparison of experiments for entangling three quantum memories with different physical systems. Notes: 1, Ref. [73] reported entanglement of two single neutral atoms, generation rate of entanglement is 0.01 counts per second. In 2017, the verification efficiency for atomic state had arrived 0.98 [110]. 2, in Ref. [75], the scheme is storage of the photons of photon-atom entanglement into another atom with EIT for generating atom-atom entanglement. The entanglement rate is higher than the work presented in this thesis. While for preparing entanglement of three single neutral atoms via three pairs of photon-atom entanglement, the three-fold coincidence rate will undergo a large decrease, especially considering loading and trapping three single atoms at the same time. 3, as an extension of Ref. [87], entanglement between two quantum dots was reported in Ref. [111] with a coincidence rate 1.2 counts per hour. It is difficult for further entangling three quantum dots. 4, in this thesis, entanglement of two cold atomic quantum memories has a generation rate 5 Hz and overall coincidence rate 0.5 Hz after considering the retrieval process. 5, in 2011, Ref. [112] reported entanglement of two macroscopic diamonds at room temperature. In 2012, Ref. [113] reported entanglement of two rare-earth ions doped crystals. Both work utilized Fock qubit. The overall coincidence rate would be very small with multiple photon detections in multi-nodes experiments

References	Physical systems	Write-out efficiency	Verification efficiency	Repetition rate (Hz)	Generation efficiency	Verification efficiency	Counting 1 h
[73]	Single neutral atom	1E-3	0.6	5E4	2.5E-10	5.4E-11	1E-2
[75]	Single neutral atom	0.12	0.009	5000	4E-4	3E-10	6E-3
[70]	Trapped ions	4E-3	$\simeq$1	4.7E5	2.13E-8	2.13E-8	36
[69]	Trapped ions	0.057	$\simeq$1	670	4.6E-5	4.6E-5	111
[79]	NV	1E-6	$\simeq$1	1E5	2.5E-19	2.5E-9	9E-11
[80]	NV	4.5E-4	$\simeq$1	2E4	2.3E-11	2.3E-11	1.6E-3
[111]	Quantum dots	1.4E-4	0.002	1.9E7	6.9E-13	5.5E-21	3.8E-10
[86]	Quantum dots	<1E-3	<0.001	2E7	<2.5E-10	<2.5E-19	<2E-8
[93]	CAEs	2.5E-3	0.15	1E4	3.9E-9	1.3E-11	5E-4
HERE	CAEs	0.02	0.3	3E5	2E-6	5E-8	6

tum networking. In Ref. [108], the authors mainly utilized three-photon continuous variables (CV) entanglement from squeezing states and stored them in three atomic ensembles with EIT protocol, this kind of CV entanglement is sensitive to photon loss and not robust to long distance photon transmission where photon loss will undergo exponential increase with increasing distance. It still needs more work to overcome the challenge. One of the replaced entangled states is the maximum entangled state, and many research groups have made great efforts to generate maximum entangled state of separate quantum memories. The first step to do this is realizing entangled state between two quantum memories served as quantum network nodes.

In 2008, Zhen-Sheng Yuan et al. generated two pairs of entanglement between photonic polarization states and spatial modes of spinwaves, which was early demonstrated by Shuai Chen et al. in 2007 [92], after interference and projective measurements of the two write-out photons, they successfully entangled the two cold atomic ensembles [93]. This experiment demonstrated the connection of adjacent nodes for quantum repeaters, it was based on two photon detection, where the influence of 'vacuum state' and photonic phase on the transmission path no more existed. It was a big step in comparison to previous work based on single photon detection. In 2012, Stephan Ritter et al. reported connection of two single neutral atom nodes [75]. They firstly generated entanglement between photonic polarization states and atomic Zeeman states through Raman process, then sent the photons to another atom for storage with EIT protocol and converted the photonic state to atomic state, thus atom-photon entanglement becomes atom-atom entanglement. At the same year, Julian Hofmann et al. used the same procedure to acquire two pairs of single atom-single photon entanglement and operated Bell-state measurements on the two photons to generate atom-atom entanglement [73]. Imam Usmanidengr et al. prepared photon pairs through spontaneous parameter down conversion (SPDC) process and used one photon as heralded signal, then used a beam splitter (BS) to split the other photon to acquire entanglement between which path the photon occurred and whether the path existed a photon, finally store the state into two rare-earth ions doped crystals, thus realize entanglement of the two crystals [113]. In 2013, H. Bernien et al. demonstrated entanglement of two NV centers via single photon detection [80]. In 2014, D. Hucul et al. realized entanglement connection between two trapped ions [70]. In 2015, Aymeric Delteil et al. reported entanglement between two separate quantum dots [111].

Although great efforts have been made to entangle two quantum network nodes, there is no one extending the number of maximum entangled quantum memories to three. One of the most important reasons is the brightness of coincident multi-photon detection. Experiments with different physical systems differ in experimental repetition rate, photon-memory generation rate, verification efficiency, etc. Blindly pursuing the overall multiple photon coincidence rate will affect the quality of entanglement. As shown in Table 1.1, I summarize the total experimental counting rate for experiments of entangling three quantum memories with state of the art reported in different physical systems. At present, the ions system presents advantages, cause it allows the verification process with near unity efficiency. Another potential system is cold atomic ensembles, although currently the photon-atom entanglement generation rate has to be kept low for suppressing the multi-excitation events, the photon-atom entanglement generation process will be deterministic by harnessing Rydberg blockade. Even with current state of the art, using the photon-atom entanglement source from cold atomic ensembles to realize the entanglement connection of three quantum network nodes, the estimated counting rate for multi-photon coincidences has also reached 6 counts per hour, enabling entangling multiple quantum network nodes. Performing three-node connection experiments is already feasible.

1.6 Thesis Structure

This thesis mainly focuses on experiments based on cold atomic ensembles, it can be divided into eight chapters as following:

- **In this chapter, Introduction**, i.e., this chapter. In this chapter, I will mainly introduce some basic concepts and recent advances.
- **Chapter** 2, **Interaction between Single Photons and Atomic Ensembles**. During constructing quantum networks, the basic step is to generate quantum correlations between photons and quantum memories, it needs more research on how to enhance the correlation. In this chapter, I will mainly introduce the theory about collective enhancement in atomic ensembles and cavity enhanced photon-atom interaction, which can guide us improve photon-atom interaction and acquire stronger photonic signal.
- **Chapter** 3, **Preparation of Cold Atomic Ensembles**. Quantum memories in this thesis are mainly based on cold atomic ensembles, so I will mainly introduce how to experimentally prepare cold atomic ensembles.
- **Chapter** 4, **Highly Retrievable Quantum Memories**. If we want to efficiently realize the connection of multiple quantum memories in a quantum network, the performance of a single quantum memory needs to be high enough, one of which is the retrieval efficiency, characterizing how efficiently the stored information can be retrieved. Therefore, in this chapter I will mainly introduce the experimental details on DLCZ quantum memory, and present how we improve the retrieval efficiency of a quantum memory.
- **Chapter** 5, **Entanglement of Three Cold Atomic Ensembles**. After acquiring efficient quantum memories, we will focus on the photon-atom entanglement fidelity, which determines the quality of final quantum networks. In this chapter, I will introduce how we experimentally generate high-fidelity photon-atom entanglement, atom-atom entanglement, and then demonstrate entanglement of three cold atomic quantum memories served as three quantum network nodes.
- **Chapter** 6, **Interference of Three Frequency Distinguished Photons**. The background for this work is presented as follows:
  On the one hand, the photon frequencies under different physical systems are usually different, and it is difficult to achieve compatibility of multiple systems. On the other hand, for the experiments on quantum supremacy such as Bose sampling, it is more realistic to prepare multiple distinguished photons. In this chapter, I will introduce an experiment of interference of multiple frequency distinguished photons from quantum memories. Our multi-photon interference experiment may lay the foundation for connecting different physical systems in the future. Meanwhile, it may also provide necessary guidance for revealing the symmetry and quantum supremacy of the interference network.
- **Chapter** 7, **Entanglement of Two Cold Atomic Ensembles via 50 km Fibers**. To connect distant quantum network nodes, in this chapter, I will introduce the experiment for entangling 50 km fibers apart quantum memories by combining efficient photon-atom entanglement and quantum frequency conversion, which

will convert near-infrared photons to telecom band photons for lower transmission loss.

- **Chapter** 8, **Summary and Outlook**.

References

1. Preskill J (1998) Lecture notes for physics 229: quantum information and computation. California Inst Technol 16
2. Bouwmeester D, Ekert A, Zeilinger A et al (2000) The physics of quantum information, vol 3. Springer, Berlin
3. Nielsen MA, Chuang I (2002) Quantum computation and quantum information
4. Acín A, Bloch I, Buhrman H, Calarco T, Eichler C, Eisert J, Esteve D, Gisin N, Glaser SJ, Jelezko F et al (2018) The quantum technologies roadmap: a european community view. New J Phys 20(8):080201
5. Gisin N, Ribordy G, Tittel W, Zbinden H (2002) Quantum cryptography. Rev Mod Phys 74(1):145
6. Kok P, Munro WJ, Nemoto K, Ralph TC, Dowling JP, Milburn GJ (2007) Linear optical quantum computing with photonic qubits. Rev Mod Phys 79(1):135
7. Scarani V, Bechmann-Pasquinucci H, Cerf NJ, Dušek M, Lütkenhaus N, Peev M (2009) The security of practical quantum key distribution. Rev Mod Phys **81**(3):1301
8. L-M Duan and Christopher Monroe (2010) Colloquium: quantum networks with trapped ions. Rev Mod Phys 82(2):1209
9. Sangouard N, Simon C, De Riedmatten H, Gisin N (2011) Quantum repeaters based on atomic ensembles and linear optics. Rev Mod Phys 83(1):33
10. Reiserer A, Rempe G (2015) Cavity-based quantum networks with single atoms and optical photons. Rev Mod Phys 87(4):1379
11. Degen CL, Reinhard F, Cappellaro P (2017) Quantum sensing. Rev Mod Phys 89(3):035002
12. Pezze L, Smerzi A, Oberthaler MK, Schmied R, Treutlein P (2018) Quantum metrology with nonclassical states of atomic ensembles. Rev Mod Phys 90(3):035005
13. Ortigoso J (2018) Twelve years before the quantum no-cloning theorem. Am J Phys 86(3):201–205
14. Kimble HJ (2008) The quantum internet. Nature 453(7198):1023
15. Simon C (2017) Towards a global quantum network. Nat Photon 11(11):678
16. Dür W, Lamprecht R, Heusler S (2017) Towards a quantum internet. Eur J Phys 38(4):043001
17. Wehner S, Elkouss D, Hanson R (2018) Quantum internet: a vision for the road ahead. Science 362(6412)
18. Jiang L, Taylor JM, Sørensen A, Lukin MD (2007) Distributed quantum computation based on small quantum registers. Phys Rev A 76(6):062323
19. Komar P, Kessler EM, Bishof M, Jiang L, Sørensen AS, Ye J, Lukin MD (2014) A quantum network of clocks. Nat Phys 10(8):582
20. Proctor TJ, Knott PA, Dunningham JA (2017) Networked quantum sensing. arXiv:1702.04271
21. Gottesman D, Jennewein T, Croke S (2012) Longer-baseline telescopes using quantum repeaters. Phys Rev Lett 109(7):070503
22. Pikovski I, Zych M, Costa F, Brukner Č (2015) Universal decoherence due to gravitational time dilation. Nat Phys 11(8):668
23. Bennett CH, Brassard G, Crépeau C, Jozsa R, Peres A, Wootters WK (1993) Teleporting an unknown quantum state via dual classical and Einstein-Podolsky-Rosen channels. Phys Rev Lett 70(13):1895
24. Prakash H, Maurya AK, Mishra MK (2012) Quantum teleportation within a quantum network. arXiv:1210.2201

25. Yin H-L, Chen T-Y, Zong-Wen Yu, Liu H, You L-X, Zhou Y-H, Chen S-J, Mao Y, Huang M-Q, Zhang W-J et al (2016) Measurement-device-independent quantum key distribution over a 404 km optical fiber. Phys Rev Lett 117(19):190501
26. Briegel H-J, Dür W, Cirac JI, Zoller P (1998) Quantum repeaters: the role of imperfect local operations in quantum communication. Phys Rev Lett 81(26):5932
27. Lvovsky AI, Sanders BC, Tittel W (2009) Optical quantum memory. Nat Photon 3(12):706
28. Simon C, Afzelius M, Appel J, de La Giroday AB, Dewhurst SJ, Gisin N, Hu CY, Jelezko F, Kröll S, Müller JH et al (2010) Quantum memories. Eur Phys J D 58(1):1–22
29. Afzelius M, Gisin N, De Riedmatten H (2015) Quantum memory for photons. Phys Today 68(12):42–47
30. Heshami K, England DG, Humphreys PC, Bustard PJ, Acosta VM, Nunn J, Sussman BJ (2016) Quantum memories: emerging applications and recent advances. J Mod Opt 63(20):2005–2028
31. Bouwmeester D, Pan J-W, Mattle K, Eibl M, Weinfurter H, Zeilinger A (1997) Experimental quantum teleportation. Nature 390(6660):575
32. Pirandola S, Eisert J, Weedbrook C, Furusawa A, Braunstein SL (2015) Advances in quantum teleportation. Nat Photon 9(10):641
33. Boschi D, Branca S, De Martini F, Hardy L, Popescu S (1998) Experimental realization of teleporting an unknown pure quantum state via dual classical and Einstein-Podolsky-Rosen channels. Phys Rev Lett 80(6):1121
34. Furusawa A, Sørensen JK, Braunstein SL, Fuchs CA, Kimble HJ, Polzik ES (1998) Unconditional quantum teleportation. Science 282(5389):706–709
35. Riebe M, Häffner H, Roos CF, Hänsel W, Benhelm J, Lancaster, Körber TW, Becher C, Schmidt-Kaler F, James DFV et al (2004) Deterministic quantum teleportation with atoms. Nature 429(6993):734
36. Barrett MD, Chiaverini J, Schaetz T, Britton J, Itano WM, Jost JD, Knill E, Langer C, Leibfried D, Ozeri R et al (2004) Deterministic quantum teleportation of atomic qubits. Nature 429(6993):737
37. Olmschenk S, Matsukevich DN, Maunz P, Hayes D, Duan L-M, Monroe C (2009) Quantum teleportation between distant matter qubits. Science 323(5913):486–489
38. Bao X-H, Xiao-Fan X, Li C-M, Yuan Z-S, Chao-Yang L, Pan J-W (2012) Quantum teleportation between remote atomic-ensemble quantum memories. Proc Nat Acad Sci 109(50):20347–20351
39. Ursin R, Jennewein T, Kofler J, Perdigues JM, Cacciapuoti L, de Matos CJ, Aspelmeyer M, Valencia A, Scheidl T, Acin A et al (2009) Space-quest, experiments with quantum entanglement in space. Europhys News 40(3):26–29
40. Ren J-G, Ping X, Yong H-L, Zhang L, Liao S-K, Yin J, Liu W-Y, Cai W-Q, Yang M, Li L et al (2017) Ground-to-satellite quantum teleportation. Nature 549(7670):70
41. Yin J, Cao Y, Li Y-H, Liao S-K, Zhang L, Ren J-G, Cai W-Q, Liu W-Y, Li B, Dai H et al (2017) Satellite-based entanglement distribution over 1200 kilometers. Science 356(6343):1140–1144
42. Liao S-K, Cai W-Q, Liu W-Y, Zhang L, Li Y, Ren J-G, Yin J, Shen Q, Cao Y, Li Z-P et al (2017) Satellite-to-ground quantum key distribution. Nature 549(7670):43–47
43. Radnaev AG (2012) Towards quantum telecommunication and a thorium nuclear clock. PhD thesis, Georgia Institute of Technology
44. Pirker A, Wallnöfer J, Dür W (2018) Modular architectures for quantum networks. New Journal of Physics 20(5):053054
45. Wallnöfer J, Zwerger M, Muschik C, Sangouard N, Dür W (2016) Two-dimensional quantum repeaters. Phys Rev A 94(5):052307
46. Appel J, Windpassinger PJ, Oblak D, Hoff UB, Kjaergaard N, Polzik ES (2009) Mesoscopic atomic entanglement for precision measurements beyond the standard quantum limit. Proc Nat Acad Sci 106(27):10960–10965
47. Xi-Lin W, Luo-Kan C, Li W, Huang H-L, Liu C, Chen C, Luo Y-H, Su Z-E, Wu D, Li Z-D, et al (2016) Experimental ten-photon entanglement. Phys Rev Lett 117(21):210502

48. Chen L-K, Li Z-D, Yao X-C, Huang M, Li W, He L, Yuan X, Zhang Y-B, Jiang X, Peng C-Z et al (2017) Observation of ten-photon entanglement using thin bib 3 o 6 crystals. Optica 4(1):77–83
49. Nunn J, Langford NK, Kolthammer WS, Champion TFM, Sprague MR, Michelberger PS, Jin X-M, England DG, Walmsley IA (2013) Enhancing multiphoton rates with quantum memories. Phys Rev Lett 110(13):133601
50. Kaneda F, Xu F, Chapman J, Kwiat PG (2017) Quantum-memory-assisted multi-photon generation for efficient quantum information processing. Optica 4(9):1034–1037
51. Alexander AL, Longdell JJ, Sellars MJ, Manson NB (2006) Photon echoes produced by switching electric fields. Phys Rev Lett 96(4):043602
52. Kraus B, Tittel W, Gisin N, Nilsson M, Kröll S, Cirac JI (2006) Quantum memory for nonstationary light fields based on controlled reversible inhomogeneous broadening. Phys Rev A 73(2):020302
53. Hétet G, Longdell JJ, Sellars MJ, Lam PK, Buchler BC (2008) Multimodal properties and dynamics of gradient echo quantum memory. Phys Rev Lett 101(20):203601
54. Hosseini M, Sparkes BM, Hétet G, Longdell JJ, Lam PK, Buchler BC (2009) Coherent optical pulse sequencer for quantum applications. Nature 461(7261):241
55. De Riedmatten H, Afzelius M, Staudt MU, Simon C, Gisin N (2008) A solid-state light–matter interface at the single-photon level. Nature 456(7223):773
56. Afzelius M, Simon C, De Riedmatten H, Gisin N (2009) Multimode quantum memory based on atomic frequency combs. Phys Rev A 79(5):052329
57. Afzelius M, Usmani I, Amari A, Lauritzen B, Walther A, Simon C, Sangouard N, Minář J, De Riedmatten H, Gisin N et al (2010) Demonstration of atomic frequency comb memory for light with spin-wave storage. Phys Rev Lett 104(4):040503
58. Harris SE (1997) Electromagnetically induced transparency. Phys Today 50(7):36–42
59. Fleischhauer M, Lukin MD (2000) Dark-state polaritons in electromagnetically induced transparency. Phys Rev Lett 84(22):5094
60. Liu C, Dutton Z, Behroozi CH, Hau LV (2001) Observation of coherent optical information storage in an atomic medium using halted light pulses. Nature 409(6819):490
61. Fleischhauer M, Imamoglu A, Marangos JP (2005) Electromagnetically induced transparency: optics in coherent media. Rev Mod Phys 77(2):633
62. Reim KF, Nunn J, Lorenz VO, Sussman BJ, Lee KC, Langford NK, Jaksch D, Walmsley IA (2010) Towards high-speed optical quantum memories. Nat Photon 4(4):218
63. Ding D-S, Zhang W, Zhou Z-Y, Shi S, Shi B-S, Guo G-C (2015) Raman quantum memory of photonic polarized entanglement. Nat Photon 9(5):332
64. Duan L-M, Lukin MD, Cirac JI, Zoller P (2001) Long-distance quantum communication with atomic ensembles and linear optics. Nature 414(6862):413
65. Cho Y-W, Campbell GT, Everett JL, Bernu J, Higginbottom DB, Cao MT, Geng J, Robins NP, Lam PK, Buchler BC (2016) Highly efficient optical quantum memory with long coherence time in cold atoms. Optica 3(1):100–107
66. Monroe C (2002) Quantum information processing with atoms and photons. Nature 416(6877):238
67. Tian L, Rabl P, Blatt R, Zoller P (2004) Interfacing quantum-optical and solid-state qubits. Phys Rev Lett 92(24):247902
68. Moehring DL, Maunz P, Olmschenk S, Younge KC, Matsukevich DN, Duan L-M, Monroe C (2007) Entanglement of single-atom quantum bits at a distance. Nature 449(7158):68
69. Stute A, Casabone B, Schindler P, Monz T, Schmidt PO, Brandstätter B, Northup TE, Blatt R (2012) Tunable ion-photon entanglement in an optical cavity. Nature 485(7399):482
70. Hucul D, Inlek IV, Vittorini G, Crocker C, Debnath S, Clark SM, Monroe Cl (2015) Modular entanglement of atomic qubits using photons and phonons. Nat Phys 11(1):37
71. Wilk T, Webster SC, Kuhn A, Rempe G (2007) Single-atom single-photon quantum interface. Science 317(5837):488–490
72. Rosenfeld W, Hocke F, Henkel F, Krug M, Volz J, Weber M, Weinfurter H (2008) Towards long-distance atom-photon entanglement. Phys Rev Lett 101(26):260403

73. Hofmann J, Krug M, Ortegel N, Gérard L, Weber M, Rosenfeld W, Weinfurter H (2012) Heralded entanglement between widely separated atoms. Science 337(6090):72–75
74. Specht HP, Nölleke C, Reiserer A, Uphoff M, Figueroa E, Ritter S, Rempe G (2011) A single-atom quantum memory. Nature 473(7346):190
75. Ritter S, Nölleke C, Hahn C, Reiserer A, Neuzner A, Uphoff M, Mücke M, Figueroa E, Bochmann J, Rempe G (2012) An elementary quantum network of single atoms in optical cavities. Nature 484(7393):195
76. Blok MS, Kalb N, Reiserer A, Taminiau TH, Hanson R (2015) Towards quantum networks of single spins: analysis of a quantum memory with an optical interface in diamond. Faraday Discuss 184:173–182
77. Reiserer A, Kalb N, Blok MS, van Bemmelen KJM, Taminiau TH, Hanson R, Twitchen DJ, Markham M (2016) Robust quantum-network memory using decoherence-protected subspaces of nuclear spins. Phys Rev X 6(2):021040
78. Sukachev DD, Sipahigil A, Nguyen CT, Bhaskar MK, Evans RE, Jelezko F, Lukin MD (2017) Silicon-vacancy spin qubit in diamond: a quantum memory exceeding 10 ms with single-shot state readout. Phys Rev Lett 119(22):223602
79. Togan E, Chu Y, Trifonov AS, Jiang L, Maze J, Childress L, Gurudev Dutt MV, Sørensen AS, Hemmer PR, Zibrov AS et al (2010) Quantum entanglement between an optical photon and a solid-state spin qubit. Nature 466(7307):730
80. Bernien H, Hensen B, Pfaff W, Koolstra G, Blok MS, Robledo L, Taminiau TH, Markham M, Twitchen DJ, Childress L et al (2013) Heralded entanglement between solid-state qubits separated by three metres. Nature 497(7447):86
81. Hedges MP, Longdell JJ, Li Y, Sellars MJ (2010) Efficient quantum memory for light. Nature 465(7301):1052
82. Afzelius M, Chaneliere T, Cone RL, Kröll S, Moiseev SA, Sellars M et al (2010) Photon-echo quantum memory in solid state systems. Laser Photon Rev 4(2):244–267
83. Afzelius M, de Riedmatten H (2015) Quantum light storage in solid state atomic ensembles engineering the atom–photon interaction: controlling fundamental processes with photons. In: Predojević A, Mitchell MW (eds) Atoms and solids (Nano-Optics and Nanophotonics). Springer, Berlin, pp 241–268
84. Seri A, Corrielli G, Lago-Rivera D, Lenhard A, de Riedmatten H, Osellame R, Mazzera M (2018) A novel integrated platform for quantum storage of heralded single photons. arXiv:1805.02494
85. Ladd TD, van Loock P, Nemoto K, Munro WJ, Yamamoto Y (2006) Hybrid quantum repeater based on dispersive CQED interactions between matter qubits and bright coherent light. New J Phys 8(9):184
86. De Greve K, Yu L, McMahon PL, Pelc JS, Natarajan CM, Kim NY, Abe E, Maier S, Schneider C, Kamp M et al (2012) Quantum-dot spin–photon entanglement via frequency downconversion to telecom wavelength. Nature 491(7424):421
87. Gao WB, Fallahi P, Togan E, Miguel-Sánchez J, Imamoglu A (2012) Observation of entanglement between a quantum dot spin and a single photon. Nature 491(7424):426
88. Tsuchimoto Y, Knüppel P, Delteil A, Sun Z, Kroner M, Imamoğlu A (2017) Proposal for a quantum interface between photonic and superconducting qubits. Phys Rev B 96(16):165312
89. Kuzmich A, Bowen WP, Boozer AD, Boca A, Chou CW, Duan L-M, Kimble HJ (2003) Generation of nonclassical photon pairs for scalable quantum communication with atomic ensembles. Nature 423(6941):731
90. Chaneliere T, Matsukevich DN, Jenkins SD, Lan S-Y, Kennedy TAB, Kuzmich A (2005) Storage and retrieval of single photons transmitted between remote quantum memories. Nature 438(7069):833
91. Simon J, Tanji H, Thompson JK, Vuletić V (2007) Interfacing collective atomic excitations and single photons. Phys Rev Lett 98(18):183601
92. Chen S, Chen Y-A, Zhao B, Yuan Z-S, Schmiedmayer J, Pan J-W (2007) Demonstration of a stable atom-photon entanglement source for quantum repeaters. Phys Rev Lett 99(18):180505

93. Yuan Z-S, Chen Y-A, Zhao B, Chen S, Schmiedmayer J, Pan J-W (2008) Experimental demonstration of a BDCZ quantum repeater node. Nature 454(7208):1098
94. Radnaev AG, Dudin YO, Zhao R, Jen HH, Jenkins SD, Kuzmich A, Kennedy TAB (2010) A quantum memory with telecom-wavelength conversion. Nat Phys 6(11):894
95. Zhao B, Chen Y-A, Bao X-H, Strassel T, Chuu C-S, Jin X-M, Schmiedmayer J, Yuan Z-S, Chen S, Pan J-W (2009) A millisecond quantum memory for scalable quantum networks. Nat Phys 5(2):95
96. Bao X-H, Reingruber A, Dietrich P, Rui J, Dück A, Strassel T, Li L, Liu N-L, Zhao B, Pan J-W (2012) Efficient and long-lived quantum memory with cold atoms inside a ring cavity. Nat Phys 8(7):517
97. Yang S-J, Wang X-J, Li J, Rui J, Bao X-H, Pan J-W (2015) Highly retrievable spin-wave-photon entanglement source. Phys Rev Lett 114(21):210501
98. Yang S-J, Wang X-J, Bao X-H, Pan J-W (2016) An efficient quantum light-matter interface with sub-second lifetime. Nat Photon 10(6):381
99. Reim KF, Michelberger P, Lee KC, Nunn J, Langford NK, Walmsley IA (2011) Single-photon-level quantum memory at room temperature. Phys Rev Lett 107(5):053603
100. Hosseini M, Sparkes BM, Campbell G, Lam PK, Buchler BC (2011) High efficiency coherent optical memory with warm rubidium vapour. Nat Commun 2:174
101. Bashkansky M, Fatemi FK, Vurgaftman I (2012) Quantum memory in warm rubidium vapor with buffer gas. Opt Lett 37(2):142–144
102. Pinel O, Hosseini M, Sparkes BM, Everett JL, Higginbottom D, Campbell GT, Lam PK, Buchler BC (2013) Gradient echo quantum memory in warm atomic vapor. J Vis Exp: JoVE 1(81)
103. Sprague MR, Michelberger PS, Champion TFM, England DG, Nunn J, Jin X-M, Kolthammer WS, Abdolvand A, Russell PSJ, Walmsley IA (2014) Broadband single-photon-level memory in a hollow-core photonic crystal fibre. Nat Photon 8(4):287
104. Sete EA, Eleuch H (2015) High-efficiency quantum state transfer and quantum memory using a mechanical oscillator. Phys Rev A 91(3):032309
105. Riedinger R, Hong S, Norte RA, Slater JA, Shang J, Krause AG, Anant V, Aspelmeyer M, Gröblacher S (2016) Non-classical correlations between single photons and phonons from a mechanical oscillator. Nature 530(7590):313
106. Riedinger R, Wallucks A, Marinković I, Löschnauer C, Aspelmeyer M, Hong S, Gröblacher S (2018) Remote quantum entanglement between two micromechanical oscillators. Nature 556(7702):473
107. Choi KS, Goban A, Papp SB, Van Enk SJ, Kimble HJ (2010) Entanglement of spin waves among four quantum memories. Nature 468(7322):412
108. Yan Z, Liang W, Jia X, Liu Y, Deng R, Li S, Wang H, Xie C, Peng K (2017) Establishing and storing of deterministic quantum entanglement among three distant atomic ensembles. Nat Commun 8(1):718
109. Chou C-W, De Riedmatten H, Felinto D, Polyakov SV, Van Enk SJ, Kimble HJ (2005) Measurement-induced entanglement for excitation stored in remote atomic ensembles. Nature 438(7069):828
110. Rosenfeld W, Burchardt D, Garthoff R, Redeker K, Ortegel N, Rau M, Weinfurter H (2017) Event-ready bell test using entangled atoms simultaneously closing detection and locality loopholes. Phys Rev Lett 119(1):010402
111. Delteil A, Sun Z, Gao W, Togan E, Faelt S, Imamoğlu A (2016) Generation of heralded entanglement between distant hole spins. Nat Phys 12(3):218
112. Lee KC, Sprague MR, Sussman BJ, Nunn J, Langford NK, Jin X-M, Champion T, Michelberger P, Reim KF, England D et al (2011) Entangling macroscopic diamonds at room temperature. Science 334(6060):1253–1256
113. Usmani I, Clausen C, Bussières F, Sangouard N, Afzelius M, Gisin N (2012) Heralded quantum entanglement between two crystals. Nat Photon 6(4):234

Chapter 2
Interaction Between Single Photons and Atomic Ensembles

Photons can be used to transmit quantum information, while atoms can be used to store and manipulate quantum information. How to achieve the interaction between the two is particularly important. In the process of constructing a quantum network, generating the correlation between photons and atoms is a very important and basic step. The theoretical analysis of how to enhance the correlation and increase the brightness of the signal photons obtained is very instructive for the experiment. Therefore, in this chapter, I will give a necessary review on the classical picture of the interaction between single photons and atoms which has been deduced in detail in Ref. [1] and the related theory in the framework of quantum mechanics [2, 3] together with some descriptions in other literatures. With this chapter, I hope I can illustrate how we can experimentally obtain high-brightness signal photons in the designated collection mode under cavity enhancement and collective enhancement in the atomic ensemble.

2.1 Photon-Atom Interaction: Classical Description

2.1.1 Photon-Atom Interaction in Free Space

2.1.1.1 Interaction Between Photons and Single Atoms

When we consider the interaction of weak light and two-level atoms (resonant frequency is ω_0, natural linewidth is Γ), the atoms can be viewed as a classical dipole oscillator [4]. For a linear polarized optical field with frequency of $\omega = ck$, it can be described as:

$$\overrightarrow{E}(t) = \frac{1}{2}\hat{e}Ee^{-i\omega t} + c.c. \tag{2.1}$$

B. Jing, *Quantum Network with Multiple Cold Atomic Ensembles*, Springer Theses,
https://doi.org/10.1007/978-981-19-0328-1_2

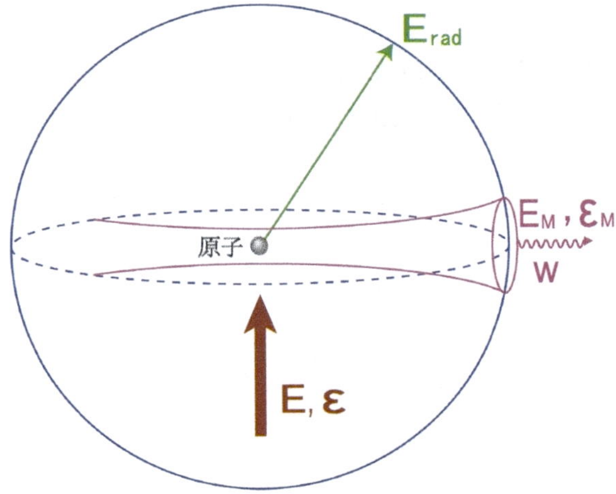

Fig. 2.1 **A schematic diagram of the light field radiating outward from atoms driven by the light field** *E*

$\hat{e}$ is the unit vector, the dipole moment of the atom induced by the light field can be then expressed as:

$$\overrightarrow{p} = \frac{1}{2}\hat{e}pe^{-i\omega t} + c.c. \tag{2.2}$$

And their complex amplitudes satisfy the relationship:

$$p = \alpha E. \tag{2.3}$$

Here α is the complex susceptibility [5]:

$$\alpha = 6\pi\epsilon_0 c^3 \frac{\Gamma/\omega_0^2}{\omega_0^2 - \omega^2 - i(\omega^3/\omega_0^2)\Gamma}. \tag{2.4}$$

This oscillating dipole will also radiate the light field outward, as shown in Fig. 2.1, the radiated optical field at a distance from the atom R and an angle of θ with $\hat{e}$ can be written as [6]:

$$E_{rad}(R,\theta) = \frac{k^2 Sin\theta}{4\pi\epsilon_0}\frac{e^{ikR}}{R}\alpha E. \tag{2.5}$$

Considering the optical field at a specific direction, the beam waist is w, electronic field is E_M, effective mode area is $A = \pi w^2/2$, we can define the amplitude of the optical field as:

$$\begin{aligned} \mathcal{E}_M &= E_{M0}\sqrt{\epsilon_0 c A}, \\ \mathcal{E} &= E\sqrt{\epsilon_0 c A}. \end{aligned} \tag{2.6}$$

By integrating E_{rad} under the effective mode area, we can get the following relationship [1].

$$\mathcal{E}_M = i\frac{k\alpha}{\pi w^2 \epsilon_0}\mathcal{E} = i\beta\mathcal{E}, \tag{2.7}$$

where $|\beta|^2 = \frac{6}{k^2w^2}\text{Im}(\beta) = \eta_{fs}\text{Im}(\beta)$. Then the cooperativity parameter between single atom and optical field in the free space can be defined and written as [1, 7]:

$$\eta_{fs} = \frac{6}{k^2w^2}. \tag{2.8}$$

By integrating E_{rad} in the whole 4π solid angle, the power of optical field from atomic radiation towards the whole space can be calculated as:

$$P_{4\pi} = \frac{ck^4}{12\pi\epsilon_0}|\alpha E|^2 = \frac{|\mathcal{E}_M|^2}{\eta_{fs}} = \frac{2P_M}{\eta_{fs}}. \tag{2.9}$$

The corresponding ratio of light power in the specific direction and the whole space can thus be acquired:

$$\frac{P_M}{P_{4\pi}} = \frac{\eta_{fs}}{2}. \tag{2.10}$$

The whole process can be understood as follows: for an incident optical field $P_{in} = \frac{1}{2}|\mathcal{E}|^2$, some is absorbed by the atom and radiated towards to the whole 4π space, the ratio is:

$$\frac{P_{4\pi}}{P_{in}} = \text{Im}(2\beta). \tag{2.11}$$

In the case of rotating-wave approximation, i.e., $\Delta \equiv \omega - \omega_0 \ll \omega_0$,

$$\beta = \eta_{fs}\left(i\frac{\Gamma^2}{\Gamma^2+4\Delta^2} + \frac{-2\Delta\Gamma}{\Gamma^2+4\Delta^2}\right). \tag{2.12}$$

Here $\mathcal{L}_a = \frac{\Gamma^2}{\Gamma^2+4\Delta^2}$ and $\mathcal{L}_d = \frac{-2\Delta\Gamma}{\Gamma^2+4\Delta^2}$ correspond to the absorption and dispersion in a Lorentzian curve. The attenuation of the incident optical field thus is:

$$\frac{P_{4\pi}}{P_{in}} = 2\eta_{fs}\mathcal{L}_a. \tag{2.13}$$

Meanwhile, the scattering cross-section of the atom can be get:

$$\sigma = \frac{P_{4\pi}}{P_{in}/A} = \text{Im}(2\beta)A. \tag{2.14}$$

In the resonant case, i.e., $\Delta = 0$, the absorption of light is related to the optical depth of the single atom, i.e.,

$$\eta_{abs} = \frac{P_{4\pi}}{P_{in}} = 1 - e^{-od}. \tag{2.15}$$

The interference between the radiated field and the transmitted field of the incident light will also result in phase shift on the light, which is also corresponded to the dispersion process. For simplicity, here we consider the radiated field collinear with the incident light, the interference will lead to a final light field as:

$$\mathcal{E} + \mathcal{E}_M = (1 + i\beta)\mathcal{E} \approx e^{i\beta}\mathcal{E}. \tag{2.16}$$

The extra phase is $\phi = \mathrm{Re}(\beta)$. By making use of Eq. 2.12, we can get the phase shift of the incident light passing through the atom as:

$$\phi = \eta_{fs}\mathcal{L}_d(\Delta). \tag{2.17}$$

During the whole process of photon-atom interaction, the radiated light field can be understood as follows: the light that is scattered by atoms or spontaneously radiated after the atoms absorb the incident light interferes with the existed light field, so there will be absorption and dispersion of light after passing through atoms.

2.1.1.2 Interaction Between Photons and Atomic Ensembles: Collective Enhancement

If it is not a single atom interacting with the optical field but an atomic ensemble with N atoms, then Eq. 2.7 becomes:

$$\mathcal{E}_M = iF\beta\mathcal{E}, \tag{2.18}$$

where F is a collective coupling coefficient related to atomic position r_j. It can be written as:

$$F = \sum_{j=1}^{N} e^{i(\boldsymbol{k}-\boldsymbol{k}_M)\cdot \boldsymbol{r}_j}. \tag{2.19}$$

Similar with the single atom case, we can get:

$$\frac{P_M}{P_{4\pi}} = \frac{1}{2}|F|^2\eta_{fs}. \tag{2.20}$$

It can be seen that the radiated field in a specific direction can change from destructive interference to collective enhanced interference due to the collective effect. Whether it is destructive or collective enhancement depends on the photon phases from different atoms. If $(\boldsymbol{k} - \boldsymbol{k}_M) \cdot \boldsymbol{r}_j$ for different atoms has opposite phases, $\langle|F|^2\rangle = 0$, the optical field gets destructive interference, no photons will be radiated in such directions. If $(\boldsymbol{k} - \boldsymbol{k}_M) \cdot \boldsymbol{r}_j$ are completely random, $\langle|F|^2\rangle = N$, in contrast to the case for single atom (Eq. 2.7), the field is improved by N times, i.e., the power of radiated field is proportional to atom number. This is actually the fact that there is

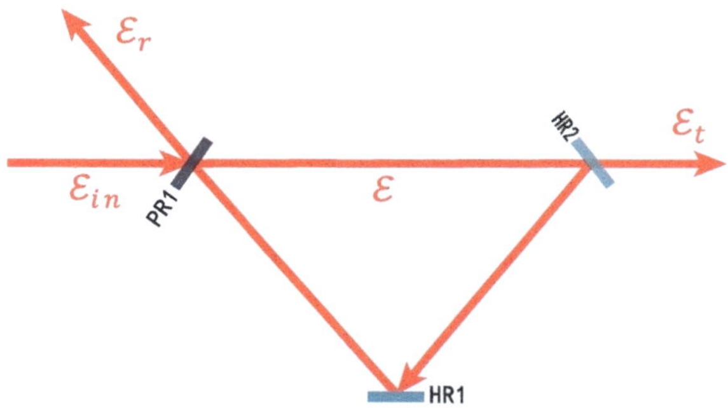

Fig. 2.2 Schematic for photons passing through a ring cavity

no influence among atoms, the emitted photon is the simple sum of the intensity of the radiated field of each individual atom. There are N atoms, which will naturally increase the field by a factor of N. If $(\boldsymbol{k}-\boldsymbol{k}_M)\cdot\boldsymbol{r}_j$ for all j are the same, $\langle|F|^2\rangle = N^2$, P_M will be enhanced by a factor of N^2 compared to single atom case, this effect is the so called collective enhancement effect. It significantly improves the probability of photons emitting toward a specific direction. In the DLCZ quantum memory, it is also based on this collective enhancement effect that the read-out photons in the DLCZ protocol can be emitted mainly along the specified direction, which greatly improves the retrieval efficiency of this type of memory.

2.1.2 *Photon-Atom Interaction Inside a Ring Cavity*

2.1.2.1 Fundamentals of a Cavity

When the light passes through the cavity, there will be some different situations from the free space. Since only a specific mode of light can exist in the cavity, the cavity will have different degrees of enhancement, weakening and mode selection on the light field, here I briefly introduces the specific situation when the light field passes through a ring cavity, as shown in Fig. 2.2. The incident light field is $\mathcal{E}_{in}$ (frequency of f, $\omega = 2\pi f$), after passing through a partial reflective mirror PR1 (reflectivity: $R_1 = r_1^2$, transmissivity: $T_1 = t_1^2$), part of the light enters the cavity while the residual part is directly reflected. The light field after reflection is $\mathcal{E}_r$, the light field inside the cavity is $\mathcal{E}$. When the light in the cavity passes through a high reflectivity mirror HR2 (reflectivity: $R_2 = r_2^2$, transmissivity: $T_2 = t_2^2$), part of the light ($\mathcal{E}_t$) escapes from the cavity. The length of the cavity is L, for simplicity, the high reflectivity mirror HR1 is assumed to have a 100% reflectivity. The light field passing through the cavity can be regarded as a comprehensive light field composed of multiple light fields coherently interfered after the light field passes through several mirrors again and again. The amplitude of the light field entering the cavity for the first time is:

$$\mathcal{E}_1 = \mathcal{E}_{in} t_1. \tag{2.21}$$

After a round trip in the cavity, the residual field is:

$$\mathcal{E}_2 = \mathcal{E}_1 t_2 r_1 e^{i\omega L/c}. \tag{2.22}$$

By analogy, we can get the situation of travelling multiple times in the cavity:

$$\begin{aligned}
\mathcal{E}_3 &= \mathcal{E}_2 t_2 r_1 e^{i\omega L/c} \\
\mathcal{E}_4 &= \mathcal{E}_3 t_2 r_1 e^{i\omega L/c} \\
&\vdots \\
\mathcal{E}_i &= \mathcal{E}_{i-1} t_2 r_1 e^{i\omega L/c} \\
&\vdots
\end{aligned} \tag{2.23}$$

The final light field is the interference of all the aforementioned field, i.e.,

$$\mathcal{E} = \sum_i \mathcal{E}_i = \frac{t_1}{1 - r_2 r_1 e^{i\omega L/c}} \mathcal{E}_{in}. \tag{2.24}$$

In another view, the light field in the same position in the cavity should keep the same, i.e.,

$$\mathcal{E} = \mathcal{E}_{in} t_1 + \mathcal{E} r_2 r_1 e^{i\omega L/c}. \tag{2.25}$$

We can also get:

$$\mathcal{E} = \frac{t_1}{1 - r_2 r_1 e^{i\omega L/c}} \mathcal{E}_{in}. \tag{2.26}$$

Similarly, we can acquire the light field $\mathcal{E}_r, \mathcal{E}_t$ after being reflected by PR1 and transmitted after HR2:

$$\begin{aligned}
\mathcal{E}_r &= \left(-r_1 + \frac{t_1^2 r_2}{1 - r_2 r_1 e^{i\omega L/c}} \right) \mathcal{E}_{in}, \\
\mathcal{E}_t &= \frac{t_1 t_2}{1 - r_2 r_1 e^{i\omega L/c}} \mathcal{E}_{in}.
\end{aligned} \tag{2.27}$$

The corresponding light intensities are:

$$\begin{aligned}
I_r &= | - r_1 + \frac{t_1^2 r_2}{1 - r_2 r_1 e^{i\omega L/c}} |^2 \mathcal{E}_{in}^2, \\
I_t &= | \frac{t_1 t_2}{1 - r_2 r_1 e^{i\omega L/c}} |^2 \mathcal{E}_{in}^2 \approx \frac{I_{max}}{1 + (\frac{\mathcal{F}}{\pi})^2 (\frac{\delta}{\nu_{FSR}})^2}.
\end{aligned} \tag{2.28}$$

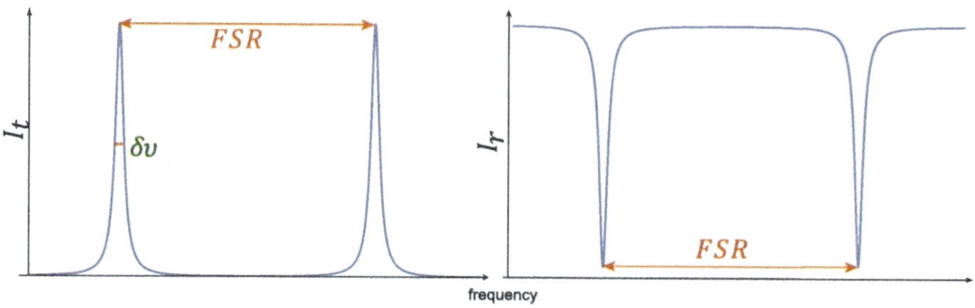

Fig. 2.3 Reflected and transmitted signal of light passing through the ring cavity

Here $I_{max} = |\frac{t_1 t_2}{1-r_2 r_1}|^2 \mathcal{E}_{in}^2$ is the intensity of transmitted light when the light is in resonance with the cavity. $\mathcal{F} = \frac{\pi\sqrt{r_1 r_2}}{1-r_1 r_2}$ is the cavity finesse, $\delta = \omega - \omega_c$ is the frequency detuning of light from the cavity resonance, $\nu_{FSR} = c/L$ is the period of the transmitted signal, also the so called free spectral range. It can be seen that for a specific cavity, as the frequency of the photon is different, the phase $\varphi = \omega L/c = 2\pi f L/c$ accumulated in the cavity will be different accordingly. The phase after and before one round trip may appear in any phase such as from the same phase to the opposite phase. In-phase means that the interference is constructive, while anti-phase means that the interference is destructive. By adjusting the frequency, the transmission peak and reflection peak signals of the light field will periodically vary, as shown in Fig. 2.3, the period is ν_{FSR}. The cavity linewidth $\delta\nu$ is defined as the full width at half maximum of the transmission signal, the cavity finesse can be calculated as $\mathcal{F} = \nu_{FSR}/\delta\nu$. It can be seen from Eq. 2.28 that the transmission peak signal of the cavity is a Lorentz line. After the inverse Fourier transform, it can be inferred that it should be exponential decay in the time domain, and the attenuation coefficient κ is just related to the line width of the corresponding cavity ($\kappa = 2\pi\delta\nu$). So we can get the average time that the photon stays in the cavity as:

$$\tau = 1/\kappa. \tag{2.29}$$

Correspondingly, the number of round trips for a photon around the cavity is:

$$n = \frac{\tau}{L/c} = \frac{\mathcal{F}}{2\pi}. \tag{2.30}$$

If there are atoms in the cavity, the light will interact with the atoms every time it travels around the cavity, which greatly improves the strength of the photon-atom interaction compared to the single photon-atom interaction.

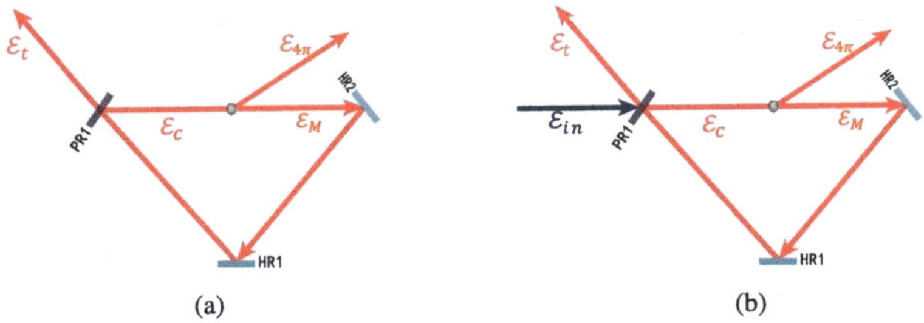

Fig. 2.4 Schematic for radiated light field from the atoms inside a ring cavity. a Schematic for radiated light field from a single atom inside a ring cavity. **b** Schematic for radiated light field from single atom inside a ring cavity with incident light field $\mathcal{E}_{in}$

2.1.2.2 Interaction Between Photons and Single Atom Inside a Cavity

As shown in Fig. 2.4a, a single two-level atom radiates light field toward the cavity mode ($\mathcal{E}_M$) while the field toward the free space outside the cavity is $\mathcal{E}_{4\pi}$. For simplicity, here we consider both HR1 and HR2 having 100% reflectivity. Only partial reflective mirror PR1 has a non-unity reflectivity ($r = r_1$) and transmissivity ($t = t_1$). For the light field inside the cavity, we have:

$$\mathcal{E}_c = re^{i\omega L/c}\mathcal{E}_c + \mathcal{E}_M. \tag{2.31}$$

If the detuning $\delta = \omega - \omega_c \ll \nu_{FSR}$, it becomes:

$$\mathcal{E}_c = \frac{\mathcal{E}_M}{\frac{1}{2}t^2 - i\frac{\delta}{\nu_{FSR}}}. \tag{2.32}$$

Through $P_t = \frac{1}{2}t^2\mathcal{E}_c^2$, $P_{4\pi} = \frac{1}{2}\mathcal{E}_{4\pi}^2$, the ratio of the power of the radiated light field along the cavity mode direction and along the other directions can be obtained as:

$$\frac{P_t}{P_{4\pi}} \approx \frac{2t^2\eta_{fs}}{t^4 + 4(\delta/\nu_{FSR})^2}. \tag{2.33}$$

When $t \to 0$, $\mathcal{F} \approx 2\pi/t^2$, let $\kappa = \nu_{FSR}t^2$, then we can get:

$$\frac{P_t}{P_{4\pi}} = \frac{2\eta_{fs}}{t^2}\frac{1}{1 + 4(\delta/\kappa)^2} = \frac{\mathcal{F}\eta_{fs}}{\pi}\frac{1}{1 + 4(\delta/\kappa)^2}. \tag{2.34}$$

If the light is in resonance with the cavity, i.e., $\delta = 0$, the equation can be simplified as:

$$\frac{P_t}{P_{4\pi}} = \frac{\mathcal{F}\eta_{fs}}{\pi}. \tag{2.35}$$

In contrast to the power of the light field radiating along the specified direction in free space (Eq. 2.10), it can be seen that with cavity enhancement the power of radiated light along the cavity mode is enhanced by a factor of $2\mathcal{F}/\pi$.

Now we consider the situation where an incident light is sent and interacts with the atom, as shown in Fig. 2.4b. The incident light ($\mathcal{E}_{in}$) is along the cavity mode direction. Similar with Eq. 2.31, we can get:

$$\mathcal{E}_c = re^{i\omega L/c}\mathcal{E}_c + t\mathcal{E}_{in} + \mathcal{E}_M. \tag{2.36}$$

By substituting $\mathcal{E}_M = i\beta\mathcal{E}_c$, we can get the light field inside the cavity as:

$$\mathcal{E}_c = \frac{\mathcal{E}_{in}/t}{\frac{1}{2} - i\frac{\delta}{\kappa} - i\frac{\beta}{t^2}}. \tag{2.37}$$

The power ratio of scattered light towards the 4π free space via the atom and incident light is:

$$\frac{P_{4\pi}}{P_{in}} = \frac{Im(8\beta)}{t^2}\frac{1}{[1 + Im(\frac{2\beta}{t^2})]^2 + [\frac{2\delta}{\kappa} + Re(\frac{2\beta}{t^2})]^2}. \tag{2.38}$$

Then we can define the cooperativity parameter with cavity enhanced photon-atom interaction as:

$$\eta = \frac{2\eta_{fs}}{t^2} = \frac{\mathcal{F}}{\pi}\eta_{fs}, \tag{2.39}$$

then we can get:

$$\frac{2\beta}{t^2} = \eta[\mathcal{L}_d(\Delta) + i\mathcal{L}_a(\Delta)], \tag{2.40}$$

which corresponds to Eq. 2.12. The atom induced phase on the photons can be also calculated as:

$$\phi = \eta\mathcal{L}_d(\Delta). \tag{2.41}$$

If the light is resonant with the atomic transition and cavity simultaneously, i.e., $\Delta = \delta = 0$, $\mathcal{L}_d = 0$, $\mathcal{L}_a = 1$.

$$\frac{P_{4\pi}}{P_{in}} = \frac{4\eta}{(1 + \eta)^2}. \tag{2.42}$$

In the case of weak coupling strength,

$$\frac{P_{4\pi}}{P_{in}} \approx 4\eta = \frac{2\mathcal{F}}{\pi} \cdot 2\eta_{fs}. \tag{2.43}$$

It represents the light absorption of the atom and the absorption is enhanced by a factor of $2\mathcal{F}/\pi$ compared to the free space case. In the case of strong coupling, for instance, with a ultra-high finesse cavity, $\eta \gg 1$,

$$\frac{P_{4\pi}}{P_{in}} \approx \frac{4}{\eta}. \tag{2.44}$$

With increasing coupling strength, $\frac{P_{4\pi}}{P_{in}} \approx 0$, the light will no longer be emitted in other directions in the whole free space but mainly towards the cavity mode direction.

2.1.2.3 Interaction Between Photons and Atomic Ensemble Inside a Cavity

Similar with the photon-atom interaction in the free space, when there exists N atoms inside the cavity, the scattered light field can be viewed as a interference of light from each atom, the collective coupling efficient is also calculated as Eq. 2.19, then we can also have:

$$\frac{P_{4\pi}}{P_{in}} = \frac{Im(8\,F|^2\beta)}{t^2}\frac{1}{[1 + Im(\frac{2\,F|^2\beta}{t^2})]^2 + [\frac{2\delta}{\kappa} + Re(\frac{2\,F|^2\beta}{t^2})]^2}. \tag{2.45}$$

2.2 Cavity Enhanced Photon-Atom Interaction: Quantum Theory

2.2.1 *Jaynes-Cummings Model*

Here we consider the evolution of a two-level atom (ground state: $|g\rangle$, excited state: $|e\rangle$) inside the cavity, according to the Jaynes-Cummings model [8], if the decay of atom and cavity is ignored, then the Hamiltonian can be written as [8–10]:

$$H_{JC}/\hbar = \omega_a\sigma_{ee} + \omega_c a^\dagger a - g(a^\dagger\sigma_{ge} + a\sigma_{eg}), \tag{2.46}$$

where ω_a and ω_c represent the frequency of atomic transition and cavity resonance, $\sigma_{ij} = |i\rangle\langle j|$ with i, j representing atomic states $|g\rangle$, $|e\rangle$. $g = \boldsymbol{d} \cdot \boldsymbol{E}$ represents the dipole interaction between photons and atoms.

If $g = 0$, the two eigen states for Eq. 2.46 is $|e, N\rangle$ and $|g, N+1\rangle$. N is the Fock state for photons. The corresponding eigen energies are $N\omega_c + \omega_a$ and$(N+1)\omega_c$.[1]

If $g \neq 0$, the eigen states for Eq. 2.46 can be solved via Schrödinger equation under dressed state picture as:

$$\begin{aligned} |+, N\rangle &= Cos\frac{\theta_N}{2}|g, N+1\rangle + Sin\frac{\theta_N}{2}|e, N\rangle, \\ |-, N\rangle &= -\,Sin\frac{\theta_N}{2}|g, N+1\rangle + Cos\frac{\theta_N}{2}|e, N\rangle. \end{aligned} \tag{2.47}$$

[1] For simplicity, let $\hbar = 1$.

Here $Tan\theta_N = -\frac{2g\sqrt{1+N}}{\delta_{ca}}$, $\delta_{ca} = \omega_c - \omega_a$. The corresponding eigen energies are:

$$E_{\pm,N} = (N+1)\omega_c + \frac{1}{2}(-\delta_{ca} \pm \sqrt{4g^2(N+1) + \delta_{ca}^2}). \tag{2.48}$$

Due to the existence of the cavity, the new eigen states include coupling photonic and atomic state. When $|\delta_{ca}| \gg |g|$, $|+, N\rangle \simeq |g, N+1\rangle$, $|-, N\rangle \simeq |e, N\rangle$.

If $|\delta_{ca}| \ll |g|$, $-g/\delta_{ca} > 0$,

$$\begin{aligned} |\pm, N\rangle &= \frac{|e, N\rangle \pm |g, N+1\rangle}{\sqrt{2}} \\ E_{\pm,N} &= (N+1)\omega_c \pm g\sqrt{N+1} \end{aligned} \tag{2.49}$$

The case of $N = 0$ corresponds to the vacuum Rabi splitting. It can be seen that when the difference of atomic transition frequency and cavity resonance frequency is small, the new eigen states include the coherent superpositions of atomic ground and excited states, and the resonant peak will be splitted into two, which will influence the choice on the frequency of read beam and finally influence the retrieval efficiency in the DLCZ quantum memory. However, if the difference is large enough, the new eigen states will not include the coherent superpositions of atomic ground and excited states. This is why we set the read beam resonant with the cavity but not resonant with the atomic transition.

When we consider the decay of atom and cavity, i.e., the natural linewidth of atomic excited states $\Gamma = \frac{\omega_0^3 d^2}{3\pi\epsilon_0\hbar c^3}$ and cavity decay rate κ, the Hamiltonian can be rewritten as [9]

$$H_{JC} = (\omega_a + i\Gamma/2)\sigma_{ee} + (\omega_c + i\kappa/2)a^\dagger a - g(a^\dagger\sigma_{ge} + a\sigma_{eg}). \tag{2.50}$$

For single atom, the eigen states in consideration of N photons under dressed state picture can be also calculated as:

$$\begin{aligned} |+, N\rangle &= Cos\frac{\theta_N}{2}|g, N+1\rangle + Sin\frac{\theta_N}{2}|e, N\rangle, \\ |-, N\rangle &= -Sin\frac{\theta_N}{2}|g, N+1\rangle + Cos\frac{\theta_N}{2}|e, N\rangle. \end{aligned} \tag{2.51}$$

Here $Tan\theta_N = -\frac{2g\sqrt{1+N}}{\delta_{ca}+i(\kappa-\Gamma)/2}$. The corresponding eigen energies are:

$$\begin{aligned} E_{\pm,N} &= (N+1)\omega_c \\ &+ \frac{1}{2}\left[-\delta_{ca} + i\frac{(2N+1)\kappa + \Gamma}{2} \pm \sqrt{4(N+1)g^2 + \left(\delta_{ca} + i\frac{\kappa - \Gamma}{2}\right)^2}\right], \end{aligned} \tag{2.52}$$

where the imaginary part is related to the dissipation and absorption parts while the real part is related to the dispersion and phase shift.

2.2.2 Photon Scattering via the Atom Inside the Cavity

Now we consider an incident light (Rabi frequency of Ω_i) detuned from the cavity resonant frequency with a detuning of $\delta = \omega - \omega_c$ and not along the cavity mode direction, it is scattered via interaction with an atom inside the cavity and the scattered photons may be along the cavity mode direction or towards other directions in the free space. At then, we can only need to rewrite Eq. 2.50 as:

$$\begin{aligned} H = &(\omega_a - \delta + i\Gamma/2)\sigma_{ee} + (\omega_c - \delta + i\kappa/2)a^\dagger a \\ &- g(a^\dagger\sigma_{ge} + a\sigma_{eg}) - \frac{\Omega_i}{2}(\sigma_{eg} + \sigma_{ge}). \end{aligned} \tag{2.53}$$

By considering $H_1 = -\frac{\Omega_i}{2}(\sigma_{eg} + \sigma_{ge})$ as a perturbation term, we can solve the photon-atom state as:

$$|\psi\rangle = c_{g,1}|g, 1\rangle + c_{e,0}|e, 0\rangle, \tag{2.54}$$

and solve the coefficients as [10]:

$$\begin{aligned} c_{g,1} &= \frac{g}{g^2 - (\delta_{ca} + \delta - i\frac{\Gamma}{2})(\delta - i\frac{\kappa}{2})}\frac{\Omega_i}{2}, \\ c_{e,0} &= \frac{\delta - i\frac{\kappa}{2}}{g^2 - (\delta_{ca} + \delta - i\frac{\Gamma}{2})(\delta - i\frac{\kappa}{2})}\frac{\Omega_i}{2}. \end{aligned} \tag{2.55}$$

Thus, the scattering rate for photons towards the free space is:

$$\Gamma_{fs} = \langle\sigma_{ee}\rangle\Gamma = |c_{e,0}|^2\Gamma, \tag{2.56}$$

while the rate for photons leaving the cavity along the cavity direction is:

$$\Gamma_{cav} = \langle a^\dagger a\rangle\kappa = |c_{g,1}|^2\kappa. \tag{2.57}$$

Let $\Gamma_0 = \frac{\Omega_i^2}{\Gamma}$ represent the photon scattering rate via the atom without the cavity, assuming the photons are resonant with both atomic transition and cavity, i.e., $\delta_{ca} = \delta = 0$, then Eqs. 2.56 and 2.57 can be written as:

$$\begin{aligned}\Gamma_{fs} &= \frac{1}{(1+\eta)^2}\Gamma_0,\\ \Gamma_{cav} &= \frac{\eta}{(1+\eta)^2}\Gamma_0.\end{aligned} \tag{2.58}$$

The probabilities of the scattered photons emitting along the cavity mode direction and emitting along any directions in the free space are as follows:

$$\begin{aligned}P_{fs} &= \frac{\Gamma_{fs}}{\Gamma_{fs}+\Gamma_{cav}} = \frac{1}{1+\eta},\\ P_{cav} &= \frac{\Gamma_{cav}}{\Gamma_{fs}+\Gamma_{cav}} = \frac{\eta}{1+\eta},\end{aligned} \tag{2.59}$$

where,

$$\eta \equiv \frac{4g^2}{\kappa\Gamma}. \tag{2.60}$$

It is the cooperativity parameter for cavity enhanced photon-atom interaction, which is the same with Eq. 2.39 as deduced as following: $g = \frac{dE}{\hbar} = d\sqrt{\omega/2\epsilon_0\hbar V}$ is the single photon Rabi frequency, for Gaussian beam with a beam waist of w_c, the mode volume inside the cavity is:

$$V = AL = \int_0^\infty dre^{-\frac{2r^2}{w_c^2}}2\pi rL = \frac{\pi w_c^2 L}{2}. \tag{2.61}$$

By making use of $\kappa = 2\pi\nu_{FSR}/\mathcal{F} = \frac{2\pi c}{L\mathcal{F}}$, natural linewidth $\Gamma = \frac{\omega_0^3 d^2}{3\pi\epsilon_0\hbar c^3}$ and resonant cases $\omega = \omega_0$, Eq. 2.60 can be rewritten as:

$$\begin{aligned}\eta =& \frac{4g^2}{\kappa\Gamma} = \frac{4d^2\omega_0}{\epsilon_0\hbar\pi w_c^2 L}\frac{L\mathcal{F}}{2\pi c}\frac{3\pi\epsilon_0\hbar c^3}{\omega_0^3 d^2}\\ &= \frac{6\mathcal{F}c^2}{\pi_c^2\omega_0^2}\\ &= \frac{6\mathcal{F}}{\pi w_c^2 k^2}\\ &= \frac{\mathcal{F}}{\pi}\eta_{fs},\end{aligned} \tag{2.62}$$

which is the same with Eq. 2.39.

2.2.3 *Interaction Between Photons and Atomic Ensembles Inside a Cavity*

If there is not a single atom inside the cavity but an atomic ensemble consisting of N atoms, then the Hamiltonian needs to sum contributions from every atom inside the cavity, i.e.,

$$H = \left(\omega_a + i\frac{\Gamma}{2}\right)\sum_j \sigma_{ee}^j + \left(\omega_c + i\frac{\kappa}{2}\right)a^\dagger a - \sum_j g(a^\dagger \sigma_{ge}^j + a\sigma_{eg}^j). \quad (2.63)$$

Here the atom-atom interaction has been ignored and the interaction between photons and each atom is viewed as equal g. By define collective ladder operator:

$$\begin{aligned} \Sigma_{eg} &= \Sigma_{ge}^\dagger = \frac{1}{\sqrt{N}}\sum_j \sigma_{eg}^j, \\ \Sigma_{ee} &= \sum_j \sigma_{ee}^j. \end{aligned} \quad (2.64)$$

Equation 2.63 can be rewritten as:

$$H = (\omega_a + i\frac{\Gamma}{2})\Sigma_{ee} + (\omega_c + i\frac{\kappa}{2})a^\dagger a - g\sqrt{N}(a^\dagger \Sigma_{ge} + a\Sigma_{eg}). \quad (2.65)$$

It is similar with the single atom case, the main difference is g has an enhancement of $\sqrt{N}$ times. The cooperativity parameter will be also enhanced by N times:

$$\eta = \frac{4Ng^2}{\kappa\Gamma}. \quad (2.66)$$

Going further, these atoms may be either in ground states or excited states, we can define collective ground states $|G\rangle$ with all atoms in ground states and collective excited states $|E\rangle^n$ with n atoms in excited states.

$$\begin{aligned} |G\rangle &= |g, g, g, \ldots, g\rangle, \\ |E\rangle^n &= (\Sigma_{eg})^n|G\rangle = \frac{1}{\sqrt{C_N^n}} \sum_{\substack{l_1,\ldots,l_n \\ l_i \neq l_j}} |g, \ldots, e^{l_1}, \ldots, e^{l_n}, \ldots, g\rangle, \end{aligned} \quad (2.67)$$

and we can get

$$\begin{aligned} \langle E^{n+1}|H|E^n\rangle &= -g\sqrt{N}a^\dagger \langle E^{n+1}|\Sigma_{ge}|E^n\rangle \\ &= -ga^\dagger\sqrt{(n+1)(N-n)}. \end{aligned} \quad (2.68)$$

It represents the transition strength from a collective state to next collective state. When $n \ll N$, for instance, $n = 0$, because of the effect of atomic ensemble, it enhances the probability of spontaneous emission, which is equivalent to the transition coefficient of spontaneous emission along a specific direction increased by $\sqrt{N}$ times. This phenomenon is called superradiance, proposed by Dicke in 1954 [11]. When all atoms are in excited states, the corresponding state is called the Dicke state.

2.3 Summary

In this chapter, I mainly cite some theoretical literature to briefly introduce some theories of the interaction between photons and atoms, which are described from the classical region and quantum level, and expand from free space to cavity enhancement. Then we focus on the collective enhancement effect and the cavity enhancement effect that will greatly enhance the brightness of the photons we collect in the specified mode, which provides some indispensable theoretical guidance for the subsequent developments of cavity-enhanced quantum memory.

References

1. Tanji-Suzuki H, Leroux ID, Schleier-Smith MH, Cetina M, Grier AT, Simon J, Vuletić V (2011) Interaction between atomic ensembles and optical resonators: classical description. In: Advances in atomic, molecular, and optical physics, vol 60. Elsevier, pp 201–237
2. Fleischhauer M, Imamoglu A, Marangos JP (2005) Electromagnetically induced transparency: optics in coherent media. Rev Mod Phys 77(2):633
3. Scully MO, Zubairy MS (1999) Quantum Opt
4. Cohen-Tannoudji C, Dupont-Roc J, Grynberg G (1998) Atom-photon interactions: basic processes and applications. In: Cohen-Tannoudji C, Dupont-Roc J, Grynberg G (eds) Atom-Photon interactions: basic processes and applications. Wiley-VCH, pp 678, Mar 1998. ISBN 0-471-29336-9
5. Milonni PW, Loudon R, Berman PR, Barnett SM (2008) Linear polarizabilities of two-and three-level atoms. Phys Rev A 77(4):043835
6. Jackson JD (2012) Classical electrodynamics. Wiley
7. Fischer T (2003) Controlling the motion of an atom in an optical cavity. PhD thesis, Technische Universität München
8. Shore BW, Knight PL (1993) The Jaynes-cummings model. J Mod Opt 40(7):1195–1238
9. Simon J (2010) Cavity QED with atomic ensembles. Harvard University
10. Tanji H (2011) Few-Photon nonlinearity with an atomic ensemble in an optical cavity. Harvard University
11. Dicke RH (1954) Coherence in spontaneous radiation processes. Phys Rev 93(1):99

Chapter 3
Preparation of Cold Atomic Ensembles

As mentioned in Chap. 1, quantum memories based on cold atomic ensembles are one of the best physical systems for constructing quantum networks. In order to realize high-performance quantum memories based on the cold atomic ensembles, a very basic premise is to prepare the cold atomic ensembles. It requires capturing the atoms in a high vacuum, reducing their speed, and finally trapping them in a specific area, thus to ensure that the number and density of the trapped atoms are sufficient for further experiments. In this chapter, I will mainly introduce the vacuum, lasers, magneto-optical trap, measurements of cold atomic density and temperature involved in the preparation of the cold atoms during the experiments.

3.1 Vacuum System

As the cold atoms are captured from background atomic gases in the vacuum, there needs to be as little as possible no other species of atoms in the background gases. Thus, it requires the entire experiment to be carried out under environment with high vacuum to reduce the influence of background impurities. As shown in Fig. 3.1, indicated are the schematic diagram of the experimental platform and the built vacuum system. The requirements for vacuum system in our experiments is lower than that in experiments with ultra-cold atomic ensembles such as Bose-Einstein Condensation, therefore, the vacuum structure is relatively simple, only necessary components such as glasscell, ion pump, vacuum gauge, feedthrough, stainless steel vacuum tubing and vacuum valves are used.

The numbered devices in Fig. 3.1 and their functions are as following:

① **Ion pump** (20L/s). It is used to maintain the high vacuum, it will always be turned on after the vacuum pressure is reduced to the required value of the experiment.

② **Vacuum Gauge**. It is used to measure the vacuum pressure and monitor the

B. Jing, *Quantum Network with Multiple Cold Atomic Ensembles*, Springer Theses,
https://doi.org/10.1007/978-981-19-0328-1_3

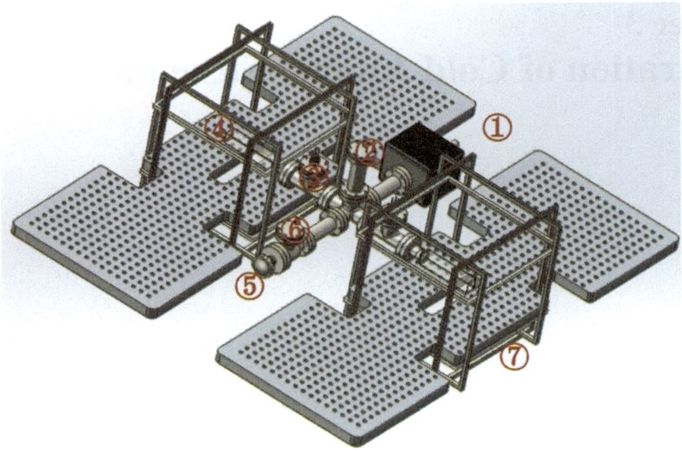

Fig. 3.1 Schematic for the platform and vacuum system

vacuum situation continuously.

③ **Feedthrough**. It is used to connect rubidium dispenser, and help generate Rb gases via applying current.

④ **Vacuum glasscell**. It is the place where Rb atoms are captured in MOT and where the experiments are carried out and interaction between atoms and photons happens. Both sides of the glasscell are coated for photonic wavelength at 795 nm, the transmission efficiency exceeds 0.99 for photons at 795 nm.

⑤ **Gate Valve**. When it is open, molecular pump and mechanical pump can be connected to ⑥ for exhausting gases in the vacuum system. After the vacuum pressure reduces to an expected value, close the gate valve and remove the molecular pump and mechanical pump, only ion pump is kept for maintain the vacuum in following experiments.

⑥ It is used to connect external pumps for pre-evacuating vacuum.

⑦ **Compensation magnetic field coil**. It is used to compensate the ambient magnetic field, so that the background magnetic field in the center of MOT can be set to zero.

During the setup of low-pressure vacuum system, we assemble the vacuum devices, connect the mechanical pump and molecular pump, open the gate valve, and perform preliminary extraction of the vacuum until the pressure no longer changes. Then we measure the vacuum pressure by the vacuum gauge, it can reach 1.0×10^{-8} mbar. At this moment, we turn on the ion pump and simultaneously start winding heating tape and baking all the vacuum devices. It is worth noticing that the upper bound of the baking temperature is mainly limited by the connection between vacuum glasscell and vacuum tube via specific glue. Temperature too high will cause the connected place to rupture. In the experiment, the baking temperature is set not higher than 110 °C. When adjusting the temperature, thermometers should be set in multiple places in the vacuum system and the temperature of each place should be slowly changed at the same time to avoid vacuum leakage caused by different temperatures in various places. With this method, the vacuum system is continued to be pumped for a few days and finally ends up staying at about 1.0×10^{-7} mbar. It is time to close all valves, decrease the temperature slowly to room temperature and use only ion pump for continue working. After several days, the vacuum system finally stays at a pressure of around 4×10^{-10} mbar.

3.2 Laser System

In our experiment, rubidium-87 atoms are the physical mediums and the two transition lines (D1 line and D2 line) of ^{87}Rb are what we mainly consider. As shown in Fig. 3.2, D1 line of ^{87}Rb is the transition between $5^2S_{1/2}$ and $5^2P_{1/2}$ with a wavelength of 794.98 nm. D2 line of ^{87}Rb is the transition between $5^2S_{1/2}$ and $5^2P_{3/2}$ with a wavelength of 780.24 nm. The separation between the two ground states ($5^2S_{1/2}$, $F = 1$ and $5^2S_{1/2}$, $F = 2$) under hyperfine structure is 6.8 GHz, if a single laser is used to generate lasers with D1 line or D2 line and laser with the other transition line is prepared through frequency shifting, the frequency shift range is too large to be efficient, which will increase the demand for the power of the laser, therefore, we utilize two lasers for each transition line to couple the two ground states and excited states. Lasers at 780 nm are self-manufactured, the methods can be found in Ref. [1], lasers at 795nm are DL100 series from Toptica company. All four lasers are semiconductor lasers with Littrow-type external cavity structure [2].

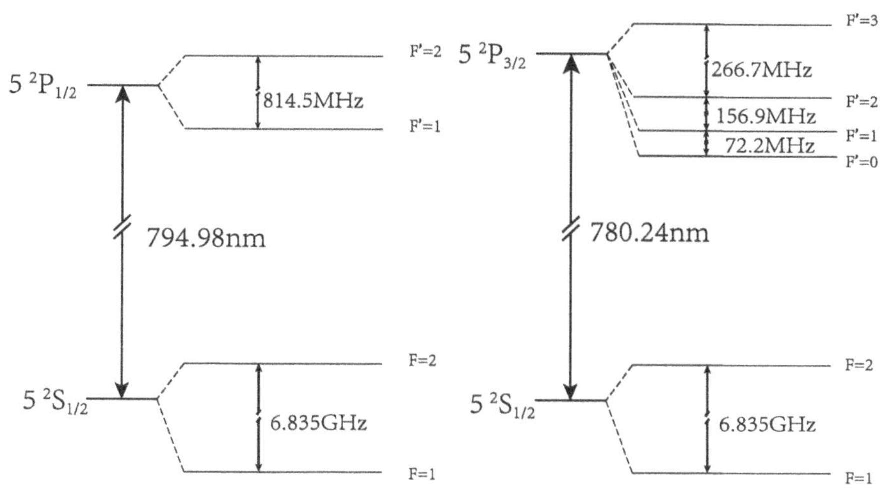

Fig. 3.2 **D1 and D2 transition line of ^{87}Rb atoms**

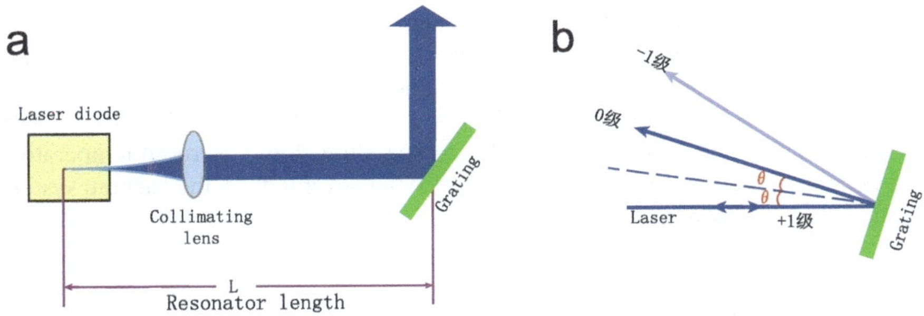

Fig. 3.3 Schematic for Littrow-type external cavity structure

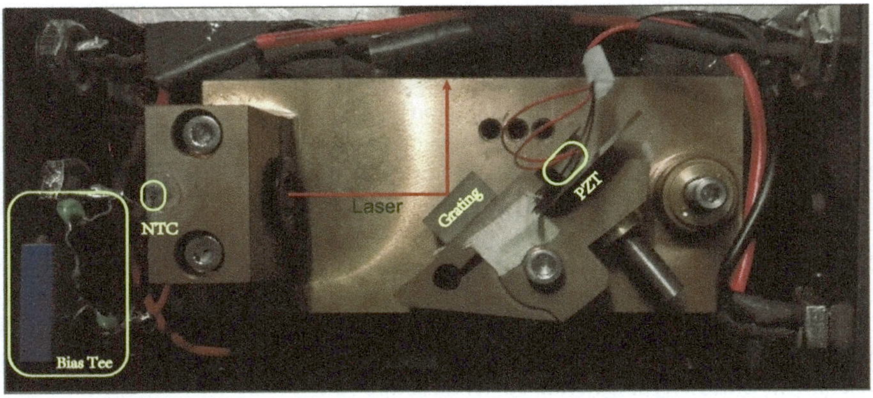

Fig. 3.4 Schematic for self-manufactured lasers at 780 nm

As shown in Fig. 3.3a, when the light emitted by a laser diode passes through a collimating lens, an external cavity is formed between the laser diode and the grating. The grating, not only narrows the laser linewidth because of the cavity, but also narrows the linewidth because of its own frequency selection. As shown in Fig. 3.3b, most of the light is diffracted to zero-order direction after the grating and is sent for the further preparation, the first-order diffracted light is reflected to the laser diode as feedback light to strengthen the resonance of the external cavity and enhance the laser output power. Using the characteristics of the blazed gratings (with a grating constant of d) and adjusting the diffracted angle θ, the wavelength of the output laser can be adjusted to meet the Bragg conditions:

$$d \sin \theta = m\lambda, \tag{3.1}$$

where m is the diffraction order, λ is the wavelength.

The grating we used for self-manufactured lasers are GH13-18V type from Thorlabs company. As shown in Fig. 3.4, the metal components in the laser devices are

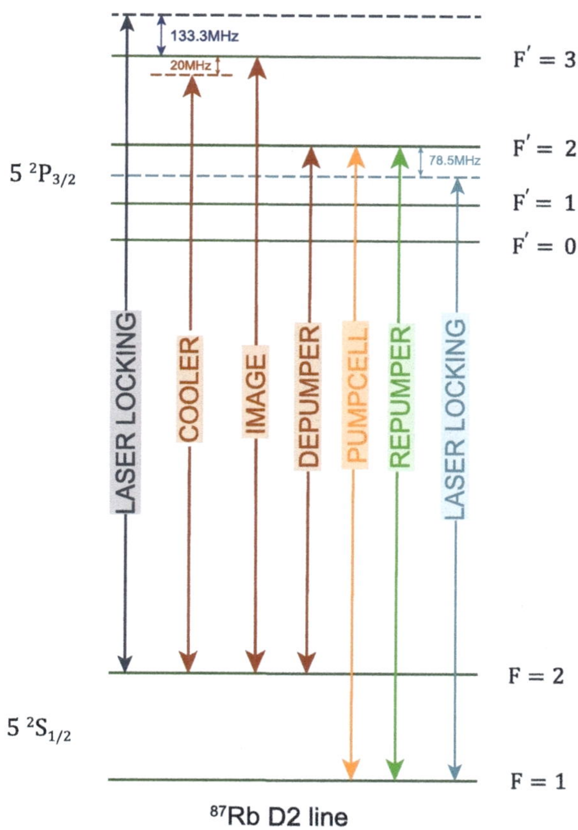

Fig. 3.5 **The frequency of each laser for the experiment with the D2 line of the rubidium-87 atom**

made of brass with good heat conduction and high hardness. There is a thermo electric cooler (TEC) underneath to control the temperature of the laser diode. A thermistor (NTC) is placed near the laser diode to monitor the temperature. A Piezoelectric Transducer (PZT) glued behind the grating is used to adjust the grating angle and feedback for locking laser frequency. Bias Tee [3] is used to couple the AC modulation signal to the laser diode to eventually generate an error signal for locking the laser frequency.

3.2.1 Lasers at 780 nm

To cool the atoms with laser cooling technology, the main lasers are used for cooler and repumper processes. Figure 3.5 shows the detailed lasers at 780 nm used during the whole experiment, the corresponding experimental setup for preparing the lasers is shown in Fig. 3.6. LASER1 and LASER2 are used for lasers related to different

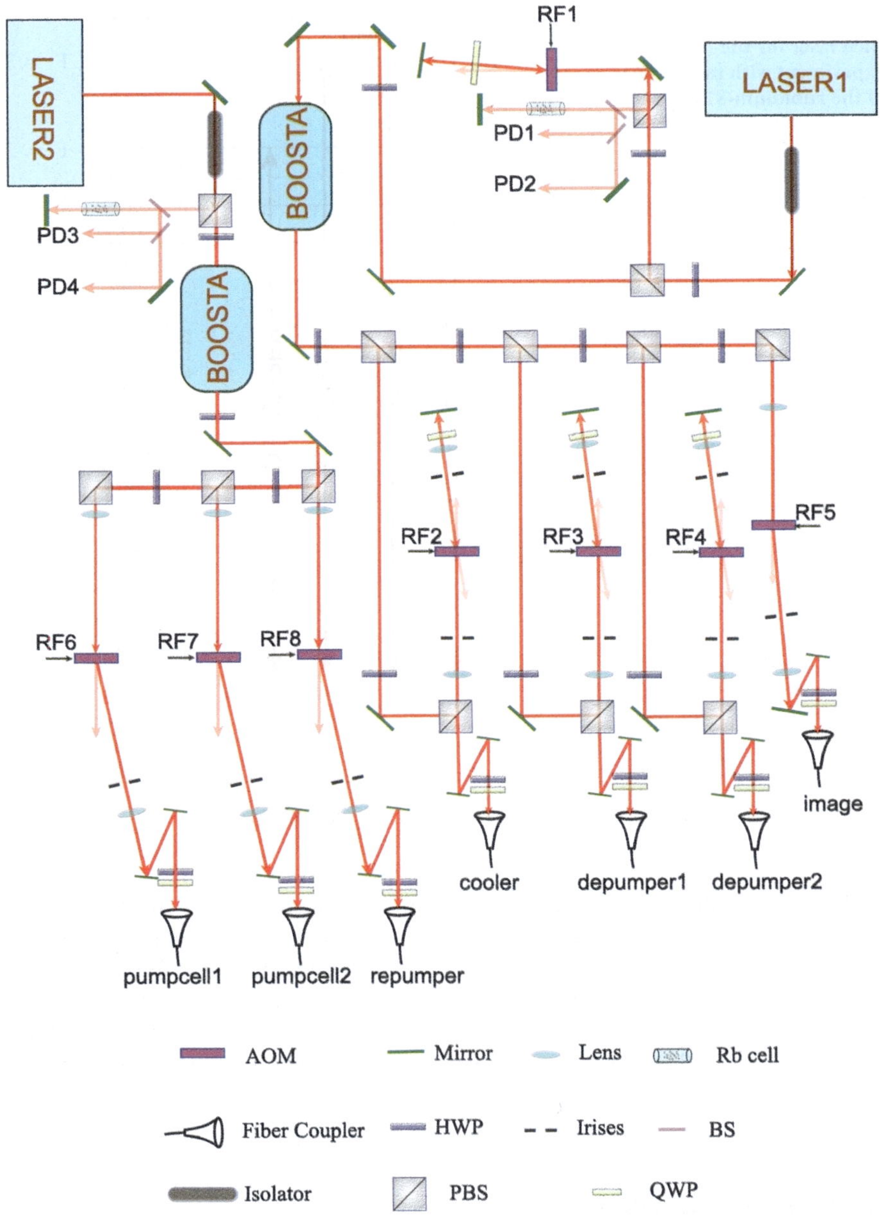

Fig. 3.6 Experimental setup for the preparation of lasers with the D2 line of rubidium-87 atom

ground states of $5^2S_{1/2}$, $F = 1$ and $5^2S_{1/2}$, $F = 2$.

LASER1 is used to generate lasers with transition of $F = 2 \rightarrow F'$, the laser beam profile from the laser diode is usually long-strip type, generally, a shaping prism or two cylindrical lenses can be used to improve the beam shape to a circle type (not shown in the figure). An optical isolator is placed after beam shaping to prevent the returned laser from injecting to the laser diode and damaging it. An AOM (acousto-optic modulator) is used to shift the laser frequency via double-pass process with a frequency shift of −266.6 MHz (RF1 = 133.3 MHz), then the laser is sent to a Rb cell for laser locking. When the voltage applied to PZT is scanned, the laser frequency is scanned correspondingly, a scanned spectrum, i.e., saturated absorption spectrum can be acquired (spectrum for D2 line of Rb is shown in Appendix A). The locking point is based on the crossover between transitions of $F = 2 \rightarrow F' = 2$ and $F = 2 \rightarrow F' = 3$, as a result, the frequency of final output laser is blue detuned from the excited state $F' = 3$ with a detuning of $\Delta_1 = 133.3$ MHz. Other lasers are split by PBSs and individually prepared with AOMs. As we will build two cold atomic ensembles, some lasers will be prepared with two sets:
1, **cooler**, AOM with double-pass, negative frequency shift (RF2 = 76.6 MHz), red detuned from $F' = 3$ with a detuning of $\Delta = -20$ MHz.
2, **depumper1**, AOM with double-pass, negative frequency shift (RF3 = 200 MHz), resonant with $F' = 2$.
3, **depumper2**, AOM with double-pass, negative frequency shift (RF4 = 200 MHz), resonant with $F' = 2$.
4, **image**, AOM with single-pass, negative frequency shift (RF5 = 133.3 MHz), resonant with $F' = 3$. In real experiments, in order to prevent the resonance light from being easily saturated to the atoms, it is often adjusted non-resonant during imaging process.

LASER2 is used to generate lasers with transition of $F = 1 \rightarrow F'$. The light from the laser directly passes through the rubidium cell to acquire the saturated absorption spectrum, based on the spectrum, the laser is locked to the crossover between transition lines of $F = 1 \rightarrow F' = 1$ and $F = 1 \rightarrow F' = 2$. The frequency of final output laser is red detuned from $F' = 2$ with a detuning of $\Delta 2 = -78.5$ MHz. Similar with LASER1, PBSs and AOMs are used to prepare the lases for different uses.
1, **repumper**, AOM with single-pass, positive frequency shift (RF6 = 78.5 MHz), resonant with $F' = 2$.
2, **pumpcell 1**, AOM with single-pass, positive frequency shift (RF7 = 78.5 MHz), resonant with $F' = 2$. The pumpcell light is used for noise filtering with atomic cells. It is mainly used to pump the atoms in the atomic cell to the required initial state so that the noise photons can be absorbed by the atoms, thus, achieving the filtering effect. After the ring cavity filter is added in the follow-up experiments, the light is not needed any more, which can only contribute to transmission losses at that time.
3, **pumpcell 2**, AOM with single-pass, positive frequency shift (RF8 = 78.5 MHz), resonant with $F' = 2$.

3.2.2 Lasers at 795 nm

The excited states of D1 line only have two splittings $F' = 1$ and $F' = 2$, less than that in D2 line, besides, the frequency separation between $F' = 1$ and $F' = 2$ is 814.5 MHz, larger than the separations in D2 line, which can get smaller crosstalk of the excited states. Because of the simpleness of excited states, it seems better to utilize D1 line for quantum memories, where only three-level Λ-structure system is used. After preparing the cold atomic ensembles, we use lasers at 795 nm to manipulate the quantum memories with DLCZ protocol. Figure 3.7 shows the 795 nm lasers used in the experiment and the corresponding frequency. Different from laser locking for D2 line, where both LASER1 and LASER2 are locked via PDH (Pound-Drever-Hall) methods [4] with saturated absorption spectrum, for LASER3 (master laser), it is used to generate lasers with transitions of $F = 1 \rightarrow F' = 2$ and locked via saturated absorption spectrum, LASER4 (slave laser) is used to generate lasers with transitions of $F = 2 \rightarrow F' = 2$ and locked via the beat signal between lasers from LASER3 and LASER4. The frequency of beat signal (ω_b) is compared to the frequency ω_0 (6.795 GHz), when $\omega_0 - \omega_b > 0$, the sign of error signal is set to "+"; when $\omega_0 - \omega_b < 0$, the sign of error signal is set to "−", when $\omega_0 - \omega_b = 0$, the error signal is set to "0". Through sending the error signal into a DP-lock device[5], it will dynamically control the laser current and PZT voltage to guarantee the slave laser having stable frequency and phase compared to the master laser. Figure 3.8 shows the saturated absorption spectrum and error signal of LASER3 and LASER4, respectively.

In consideration of laser locking and the preparation of lasers with different frequencies, the experimental details are shown in Fig. 3.9. Laser from LASER3 firstly passes through an AOM with double-pass positive frequency shift (RF9 = 125 MHz), then passes through the Rb cell and is detected by photo detectors. By scanning the voltage of PZT, i.e., scanning the laser frequency, the saturated absorption spectrum and its related error signal can be acquired (Fig. 3.8a), where the peak in the spectrum with transition from $F = 1$ to $F' = 2$ is selected as reference of the final locking point. As a result, the lasers from LASER3 has a stable frequency with 250 MHz red detuned from the excited state $F' = 2$. Laser from LASER4 is combined with laser from LASER3 at a beam splitter for interference, the beat signal from the interference is detected by photon detector PD6 and sent to the DP-lock device for locking the laser. For the convenience of monitoring, we also build an optical path (RF10 = 105MHz) to obtain the saturated absorption spectrum, which is displayed on the oscilloscope for comparison and monitoring together with the error signal.

The lasers at 795 nm is prepared according to following procedures:

1, pump1/pump2 is used to initialize the atomic states via coupling transition of $F = 1 \rightarrow F' = 1$, its polarization can be usually adjusted depending on the wanted atomic states and correspondingly named as π-pump or σ-pump laser. It is prepared by applying 564 MHz negative frequency shift to the lasers output from LASE3, where the frequency shift is operated by an AOM (RF11 = RF12 = 282 MHz) and double-pass propagation.

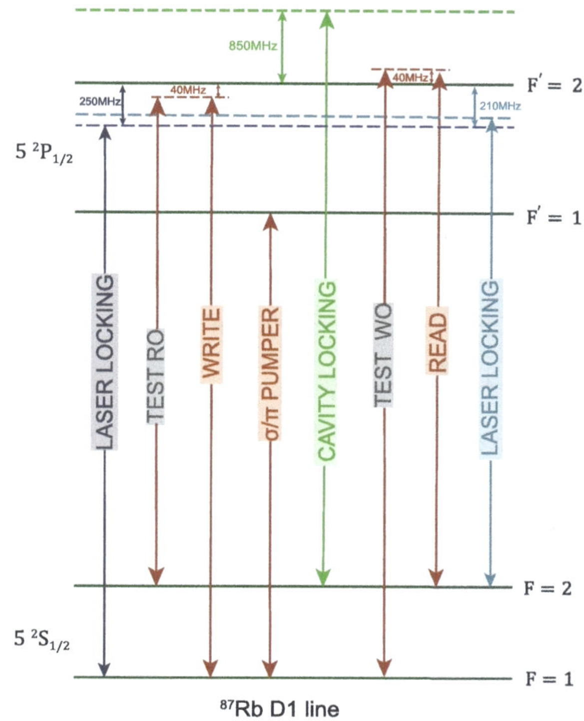

Fig. 3.7 The frequency of each laser for the experiment with the D1 line of the ^{87}Rb atom

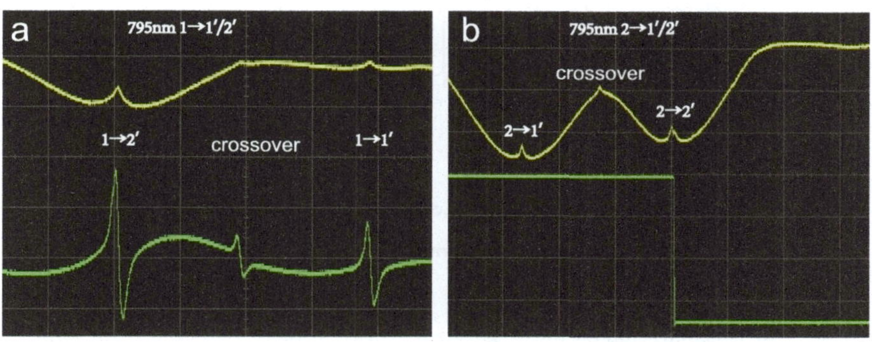

Fig. 3.8 Saturated absorption spectrum and the error signal of ^{87}Rb D1 line

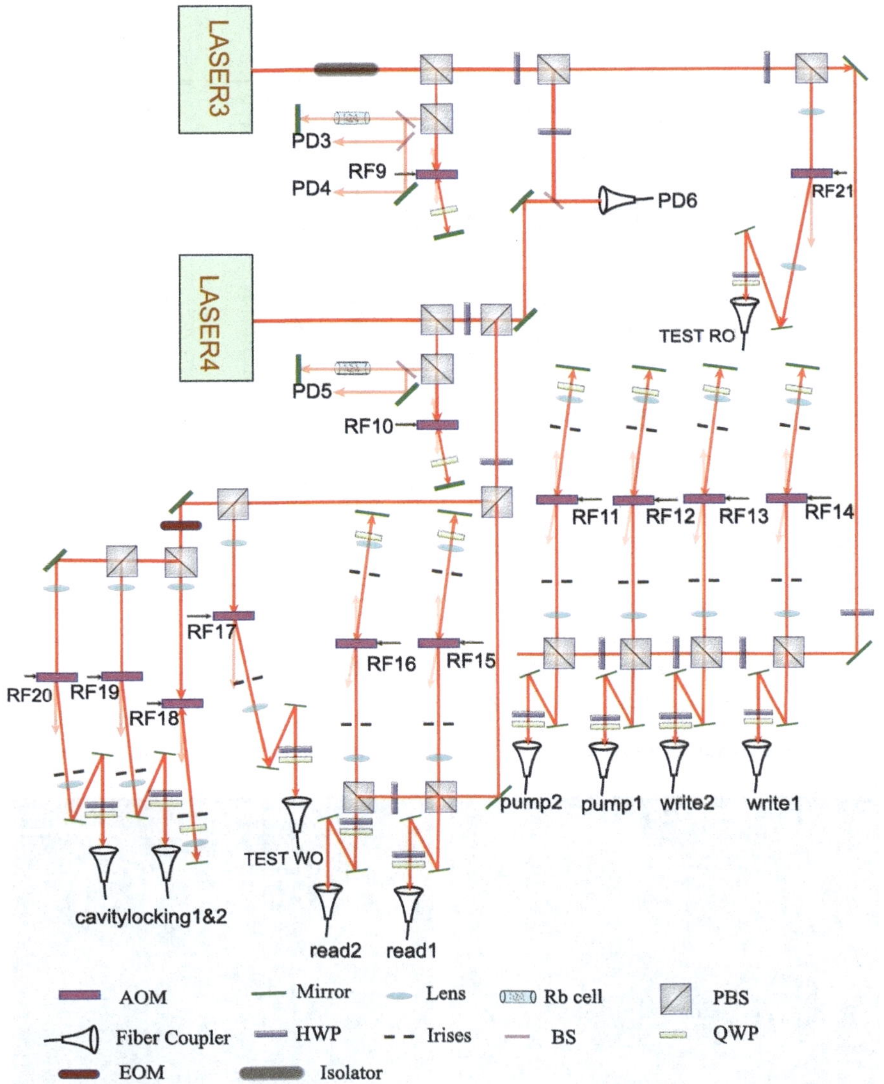

Fig. 3.9 Experimental setup for the preparation of lasers with the D1 line of ^{87}Rb atom

2, write1/write2 is the so called write beam in the DLCZ quantum memories. In the free space experiments, it is prepared by 240 MHz positive frequency shift via an AOM (RF13 = RF14 = 120 MHz) and double-pass propagation. It is finally 10 MHz red detuned from $F' = 2$. In the cavity enhanced experiments, RF13 = RF14 = 105 MHz, the detuning is set to 40 MHz.
3, TEST RO is used as the read-out photons for the adjustment of the coupling between the cavity and the read-out photons and prepared via an AOM (RF21 = 290 MHz) and single-pass propagation. Its frequency is the same with that of read-out photons and blue detuned from $F' = 2$ with a detuning of 40 MHz.
4, read1/read2 is the so called read beam in the quantum memories, it is generated from LASER4 and positive frequency shifted via an AOM (RF15 = RF16 = 125 MHz) and double-pass propagation. It is finally blue detuned from $F' = 2$ with a detuning of 40 MHz in the cavity-assisted quantum memories. While in the free space experiment, the read laser is resonant with the transition of $F = 2 \rightarrow F' = 2$.
5, TEST WO is used as the write-out photons for the adjustment of the coupling between the cavity and the write-out photons and prepared via an AOM (RF8 = 170 MHz) and single-pass propagation. Its frequency is the same with that of write-out photons and red detuned from $F' = 2$ with a detuning of 40 MHz. TEST WO and TEST RO are simultaneously resonant with the cavity.
6, cavity locking beam is used to lock the cavity, as the laser will pass through the cold atoms, it need to be far off resonant with the atomic transitions while keeping resonant with the cavity together with write-out and read-out photons. So the frequency difference between the cavity locking beam and write-out photons should be separated at least one free spectral range of the cavity. In our experiment, as we used lasers with high-order beam profiles for locking the cavity, it is generated from LASER4 and positive frequency shifted by two AOMs (RF18 = 272 MHz, RF19 = RF20 = 298 MHz) and both double-pass and single-pass propagations.

3.3 Magnetic Field

In the cold atom experiment, the existence of the background magnetic field has a vital influence on the atomic trapping and followed experiments. In our experiment, because the geomagnetic field has a magnitude of several hundred milli-gauss, it needs to be compensated to zero in the center of atomic trapping. Here we use three pairs of Helmholtz coils to generate uniform magnetic fields with opposite directions of the geomagnetic field at the place where the atoms are captured to compensate the influence of the surrounding geomagnetic field. An additional pair of Helmholtz coils are used to generate a bias magnetic field in the experiment to define the quantum magnetic axis. In the experiment, we used square Helmholtz coils to generate a uniform magnetic field in the atomic center. For a pair of coils with a side length of $2l$ and a spacing of $2a$, each coil has N turns, and the current is I. Define the plane of a coil as the $x - y$ plane, the direction of the vertical plane is the z direction, and the origin is the center of the two square coils (see Appendix B). For the places on the z axis near the center, where the magnetic field direction is along the z axis, there is no component in the $x - y$ direction, and its magnitude can be calculated as

$$\begin{aligned} B(z) &= B_z(x=0, y=0, z) \\ &= \frac{2\mu_0 N I l^2}{\pi[l^2+(z+a)^2]\sqrt{2l^2+(z+a)^2}} + \frac{2\mu_0 N I l^2}{\pi[l^2+(z-a)^2]\sqrt{2l^2+(z-a)^2}}. \end{aligned} \tag{3.2}$$

In order to have a good uniformity of the magnetic field generated in the center, this means (here only the second order is considered):

$$\begin{aligned} \frac{dB(z)}{dz}|_{z=0} &= 0 \\ \frac{d^2B(z)}{dz^2}|_{z=0} &= 0. \end{aligned} \tag{3.3}$$

It is deduced from Eq. 3.2 that the magnitude of the magnetic field is symmetric about the origin on z axis, so its first-order differentiation is zero and the second-order differentiation should be adjusted to zero, i.e., $\frac{d^2B(z)}{dz^2}|_{z=0} = 0$. By solving it, it can be calculated that when

$$a = 0.5445l, \tag{3.4}$$

$\frac{d^2B(z)}{dz^2}|_{z=0} = 0$, the uniformity of the magnetic field at the atomic trapping center is the best. However, in the real experiment, due to the limitation of the experimental space, the distance between the compensation magnetic field coils does not strictly satisfy Eq. 3.3, its side length is as long as 40 cm, the magnetic field at the center is around the atomic size (no more than 1 mm). The difference is so small that it can still be ignored. For the bias coil, considering the specific experimental conditions, we make and fix the coils near the vacuum chamber with 11 cm side length, 5 cm separation, and 12 turns for each.

3.4 Magneto-Optical Trap

3.4.1 Fundamentals

To realize coherent manipulation of cold atoms, the key step is to acquire atoms with low velocity. In 1975, Hänsch et al. proposed to cool neutral atoms with laser cooling [6], whose main idea is to utilize Doppler effect. The simple description is that when the direction of laser propagation is opposite to the direction of atomic motion, the frequency of the laser actually experienced by the atom is higher than the real frequency due to the Doppler effect. By setting a laser that is lower than the

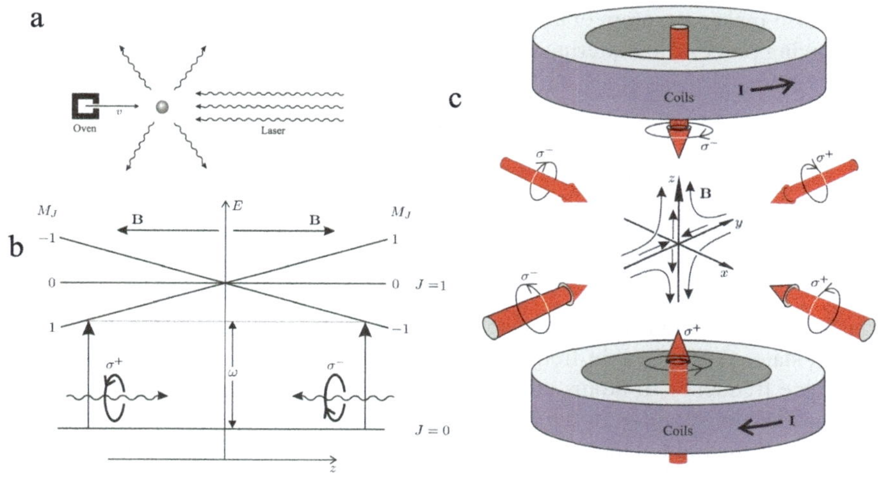

Fig. 3.10 Magneto-optical trap. Figures are reproduced from Ref. [9]

resonance frequency of the atom, the atom can absorb photons and enter an excited state if the direction of movement of the atom is opposite to the direction of the laser. When the atom falls from the excited state to the ground state, the emitted photon energy is slightly larger than that of the previous absorption, causing the atom to lose energy and slow down. Since the momentum direction of the incident photon is constant, while the angular distribution of the momentum of the emitted photon is uniform along the entire 4π space, the momentum of the atom can also be changed. Motivated by the idea, in 1985, the groups leaded by William D. Phillips and Steven Chu achieved the cooling and trapping of atoms for the first time in real experiments [7, 8], reaching the lowest temperature record of 240 μK for cold atoms at that time. Since then, more and more laser cooling technologies (such as polarization gradient cooling, Raman cooling, sideband cooling, and evaporative cooling, etc.) have achieved unprecedented developments. Up to now, atomic cooling of the order of pK has been achieved. Due to their contributions to the laser cooling technology, Steven Chu, Claude Cohen-Tannoudji and William D. Phillips shared the 1997 Nobel Prize. In our experiments, we use Doppler cooling of atoms in a gradient magnetic field to obtain low-speed cold atoms so that they can stay within a certain range during our experiments. It is the magneto-optical trap (MOT) technology.

Figure 3.10 shows the principle and schematic diagram of the magneto-optical trap (figure is from the book [9]). In Fig. 3.10a, we consider a two-level atom with ground state g and excited state e with frequency difference of ω_0, because of Doppler effect, atoms with velocity υ absorbs the laser photons(ω) below the resonance frequency, and then spontaneously radiates to the whole space. The frequency of the emitted photon is higher than the frequency of the absorbed photon, making the atom lose energy and slow down. In the aspect of momentum, atoms absorb the momentum from the opposite direction, but the average momentum during radiation is zero, which also makes the atom obtain a reverse net momentum and slow down. During the slowing process, a scattering force received by atoms can be expressed as [9]:

$$F_{sc} = pR_{sc}. \tag{3.5}$$

$p = \hbar k$ is the photon's momentum, the scattering rate is:

$$R_{sc} = \Gamma\rho_{ee} = \frac{\Gamma}{2}\frac{\Omega^2/2}{\delta^2 + \Omega^2/2 + \Gamma^2/4}, \tag{3.6}$$

where Γ is the spontaneous decay rate, ρ_{ee} refers to the population of the atom in the excited state, Ω is the rabbi frequency, $\delta = \omega - \omega_0 + kv$ is the detuning from the resonance frequency of atoms after considering the Doppler effect. Going further, we consider the situation in the magneto-optical trap, as shown in Fig. 3.10b, it is the principle diagram of the one-dimensional magneto-optical trap. Along the z axis, there is a gradient magnetic field B generated by a pair of anti-Helmholtz coil. Due to Zeeman splitting, different excited states of atoms ($M_J = 0, \pm 1$) have different energy level shifts under different position-dependent magnetic fields, the energy can be written as:

$$\Delta E(z) = g_J \mu_B M_J B(z), \tag{3.7}$$

where g_J is the Landé factor and μ_B is the Bohr magneton. For $M_J = 1$, during the process of atomic movement from $z < 0$ to $z > 0$, the energy levels of the atoms are getting higher and higher due to the different magnetic fields, while for $M_J = -1$, the opposite is true, this sub-energy level will get lower and lower, and for $M_J = 0$, it has nothing to do with the amplitude of the magnetic field, and remains unchanged throughout. Let the atoms locate in the two oppositely polarized laser beams, the laser frequency is lower than the resonance frequency of the atoms. Consider the atom moving in the direction of $z > 0$, on the one hand, it will be affected by the Doppler effect and atoms at state $M_J = -1$ is more easier to absorb the photons along the -z direction, thus obtaining the net momentum in the $-z$ direction. Atoms will slow and cool down. On the other hand, the laser along the $+z$ direction will be more difficult to interact with the atom and merely affect the atoms at state $M_J = -1$. The scattering force received by the atoms in the process can be expressed as:

$$\begin{aligned} F_{sc} &= F_+ + F_- \\ &= \frac{\hbar k\Gamma}{2}\left[\frac{\Omega^2/2}{(\omega-\omega_0+kv)^2+\Omega^2/2+\Gamma^2/4} - \frac{\Omega^2/2}{(\omega-\omega_0-kv)^2+\Omega^2/2+\Gamma^2/4}\right]. \end{aligned} \tag{3.8}$$

When the atoms has been slowed down and satisfies $|kv| \ll \Gamma$, we perform a first-order expansion near the velocity of zero, the above formula can be rewritten in the form of the damping force on the atom:

$$F_{sc} = -\alpha \upsilon, \tag{3.9}$$

where the damping coefficient α is:

$$\alpha = -\hbar k^2 \Omega^2 \frac{\Gamma(\omega - \omega_0)}{[(\omega - \omega_0)^2 + (\Omega^2/2 + \Gamma^2/4)^2]^2}. \tag{3.10}$$

Similarly, when the atom moves along the $-z$ direction, atoms at $M_J = +1$ are more likely to absorb the photons along the $+z$ direction and be cooled down. In the whole process, the continuous pumping of atoms will make these atomic states transfer to each other, so that atoms in all states can participate in the cooling process. Taken together, the atom will always obtain momentum in the opposite direction and slow down, eventually approach $z = 0$ at a very low temperature. As a result, the atoms can be trapped in the center of the magnetic field under the influence of the magnetic field. In this process, the moving atom is affected by the oppositely propagating laser, and the probability of slowing down due to the scattering force is very high while the laser beam in the same direction cannot slow it down. In this way, no matter what direction the atom is moving, it can always interact with the laser from one of the directions and slows down, i.e., be cooled down.

Theoretically, Doppler cooling can be used to cool the temperature of atoms from room temperature to μK or even nK, but there is a Doppler limit for temperature T_D and velocity υ_D. When the atom is cooled gradually, the laser and the atom no longer resonates, and the atom will no longer absorb the photons and slow down. When the movement of the atom is already very slow, the required light frequency ω will be very close to the resonance frequency ω_0 of the atom. Since the energy level of the atom has a lifetime, the corresponding natural line width determines the width of the energy level transition. Therefore, in the extreme case of this cooling method, the energy corresponding to the natural half-width $\Gamma/2$ can only be cooled, namely:

$$2 \cdot \frac{1}{2} k_b T_D = \frac{1}{2} m \upsilon_D^2 = \frac{1}{2} \hbar \Gamma. \tag{3.11}$$

Under this situation, the Doppler limit for velocity under the Doppler cooling condition can be expressed as:

$$\upsilon_D = \sqrt{\frac{\hbar \Gamma}{m}}, \tag{3.12}$$

and the Doppler limit for temperature is:

$$T_D = \frac{\hbar \Gamma}{2 k_b}. \tag{3.13}$$

We can also consider that even the recoil momentum received by a stationary atom after emitting a photon will accelerate the atom, the recoil energy obtained by the atom in this process is:

$$E_R = \frac{\hbar^2 k^2}{2m} = \frac{1}{2} k_b T_R = \frac{1}{2} m \upsilon_R^2. \tag{3.14}$$

Similarly, the recoil limit velocity that can be obtained by laser cooling can be obtained as:

$$\upsilon_R = \frac{\hbar k}{m}. \tag{3.15}$$

The recoil limit temperature is:

$$T_R = \frac{\hbar^2 k^2}{k_b m}. \tag{3.16}$$

Consider the ^{87}Rb atoms we used in the experiments, $\Gamma = 2\pi \times 6\,\text{MHz}$, atom mass with $m = 1.44 \times 10^{-25}$ kg, the corresponding Doppler limit for temperature is $T_D = 143\,\mu\text{K}$ and the recoil limit temperature is $T_R = 0.36\,\mu\text{K}$.

The principle of the above-mentioned one-dimensional case is also applicable to the three-dimensional magneto-optical trap. As shown in Fig. 3.10c, in order to cool and trap atoms in the entire three-dimensional space, three pairs of direction-opposite lasers move from east to west, north to south, and up and down. Irradiated in three directions, the laser overlapping center completely overlaps with the center of the gradient magnetic field, and the cold atom is finally captured and trapped in the center. Since the real atom is not a simple two-level atom, as shown in Fig. 3.5, in order to ensure the atoms to continuously participate in the cooling process, it is necessary to ensure that the atom is always at a specific energy level that can interact with lasers. This requires the existence of a cycle transition line, so that the atoms in a specific state can only return to the original state after absorbing light and then radiating light, thereby repeatedly participating in the Doppler cooling process to cool the atoms. Obviously, $F = 2 \rightarrow F' = 3$ satisfies such a cycle transition, so the cooling laser is fixed to couple the $F = 2 \rightarrow F' = 3$ transition, and red detuned from $F' = 3$ by 20 MHz. Even so, because of not all atoms initially in states of $F = 2$, the collision between atoms and lasers coupling other energy levels in a detuning situation, there is a certain probability that the atoms will be in the $F = 1$ state. In this way, the atoms no longer interact with the cooler laser and exit the cooling process, which is not conducive to the preparation of the cold atomic ensembles. Therefore, in order to ensure that the atoms always participate in the cooling process, a laser beam resonant with $F = 1 \rightarrow F' = 2$ is also used. The laser and the cooler laser are coupled together to ensure that atoms falling into the $F = 1$ state can be re-pumped into the $F = 2$ state. This process is called the re-pumping process, and the corresponding laser is called the re-pumped laser (Repumper).

In order to acquire more cold atoms, for the atoms that have been cooled to a very low temperature at the center, we hope that they will no longer interact with lasers and be lost, the dark magneto-optical trap technology is often used [10]. Simply put, it is to block the repumper light center in space, so that the atoms cooled in the center can keep after falling into the $F = 1$ state, instead of continuing to participate in the

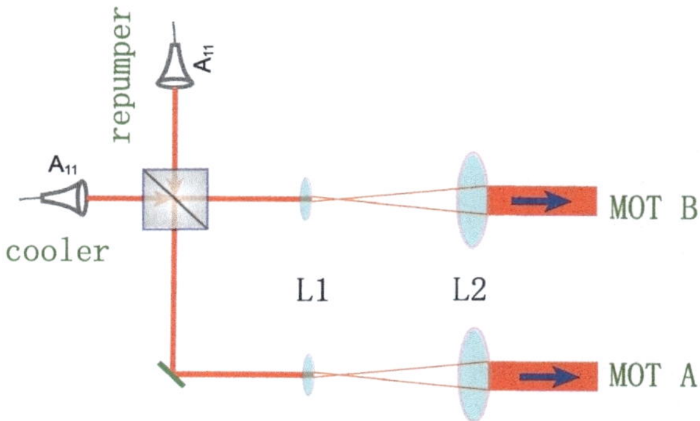

Fig. 3.11 Experimental setup for expanding cooler and repumper laser

Fig. 3.12 Cold atoms in a magneto-optical trap as seen by an infrared camera

interaction of light. It can also slowly weak the intensity of repumper laser in time to achieve a similar effect.

In order to increase the area and intensity of laser interacting with atoms, we need to ensure the beam profile and intensity of the cooler laser. In the experiment, we use the optical fiber to collect the cooler and repumper laser and use the fiber collimators (60FC-4-A11-02) produced by Schäfter+Kirchhoff to release the cooler and repumper laser, each with a power of about 90 mW and 18 mW, then the two lasers are combined into one by using a PBS. The beam waist radius of the beam spot is about 1.1 mm, later, a convex lens L1 with a focal length of $f_1 = 25$ mm (to make full use of space, a concave lens can be used here) and a convex lens L2 with $f_2 = 300$ mm are used to expand the beam into a spot with a beam waist radius of 13 mm (Fig. 3.11), and finally we divide it into six ways and send it into the vacuum chamber. Due to the limitation of the size of the waveplates used in the optical path, the passing diameter is only 1 in. (approximately 25.4 mm), so the aforementioned laser actually only enters the vacuum glasscell with a width of 25.4 mm in the middle. In practical applications, since the laser intensity distribution of Gaussian profile is not uniform, if the laser power is sufficient, it is better to enlarge the size of the beam, and then use mechanical devices to intercept the laser that needs to be used in the middle part, because the uniformity of laser intensity is better than the uniformity of the entire Gaussian light, which provides favorable conditions for ensuring that the intensity of the interaction between atoms and light in different places is similar, and the performance of cold atoms that can be obtained is also better.

In the process of adjusting the magneto-optical trap, it can be usually adjusted according to the following steps:
1, use a magnetometer to roughly measure the amplitude of the central environmental magnetic field, and to compensate the environmental magnetic field by adjusting the current of the compensating magnetic field coils to generate the magnetic field, so that the central magnetic field can be roughly adjusted to zero.
2, because a key point of preparing cold atoms is that the 6 beams of cooler laser can overlap at one point, so that the cold atoms can be captured in the center. Therefore, when adjusting the center of the intersection of the 6 lasers to overlap with the center of the magnetic field, one can use the geometric structure of the coils that generates the gradient magnetic field to adjust the 6 beams to pass through the center of the magnetic field. The two opposite lasers can be usually adjusted by setting one as the reference and adjusting the other one overlapping. In addition, one can also adjust the cooling laser to resonant with the atoms and use the irises to let only the center part of the lasers pass through, then increase the dispenser current to increase the rubidium atomic density in the vacuum, so that the laser passing through the atoms enough excite the atoms and fluorescence is emitted to indicate the direction of the lasers, similar to the several laser beams presented in Fig. 3.12. Finally, the six laser beams can be adjusted in the direction of the visible fluorescence to overlap at one point.
3, calibrate the polarizations of the six laser beams to ensure that they match with the gradient magnetic field.

The cold atoms in the completed magneto-optical trap are observed with an infrared camera. As shown in Fig. 3.12, the bright spot at the intersection of the six beams is the prepared cold atomic ensemble. When the atomic ensemble is large enough, a bright spot can be seen sporadically with the naked eyes in principle.

3.4.2 Numbers of Atoms in MOT

For the built magneto-optical trap, we hope that cold atoms can be captured in the vacuum chamber as much as possible. The number of atoms will directly affect the strength of the interaction with lasers, therefore, it is particularly important to increase the number of cold atoms. For this reason, we need to analyze what experimental parameters will affect the captured atoms.

MOT captures atoms from the background gases in the vacuum, the thermal velocity of these background gas atoms is:

$$u = \sqrt{\frac{2k_b T_{bg}}{m}}. \tag{3.17}$$

Here the temperature of background gases is room temperature. For the Rb atoms at a room temperature of 293 K, the thermal velocity is calculated as 237 m/s. Assuming there exists a velocity υ_c, atoms with velocity lower than it can be cooled down within the atom-laser interaction area, i.e., with a beam width of d, atoms cooled down can be loaded into the MOT with a loading rate as:

$$R \simeq \frac{n_{bg} d^2 \upsilon_c^4}{u^3}. \tag{3.18}$$

During the loading process, atoms absorb the photons and then spontaneously radiate, they will get a recoil velocity in each time of absorption and radiation, i.e., $\upsilon_R =$ 5.8845 m/s [11], the scattering rate is:

$$r = \frac{p\Gamma}{2(1+p)}, \tag{3.19}$$

where $p = \frac{I/I_s}{1+4\Delta^2/\Gamma^2}$ is the saturation parameter, I_s is saturation laser intensity. Therefore, the beam width of lasers that can cool down the atoms with velocity of υ_c should be larger than:

$$d = \frac{\upsilon_c^2}{2r\upsilon_R}. \tag{3.20}$$

Atoms loaded into the magneto-optical trap may collide with background atoms and escape from the MOT, the escape rate is:

$$\Gamma_c = n\sigma u. \tag{3.21}$$

σ is the scattering cross section and related to υ_c [12], finally, the number of atoms in the magneto-optical trap reaches a dynamic equilibrium, i.e.:

$$\frac{\partial N}{\partial t} = R - N\Gamma_c = 0. \tag{3.22}$$

We can now deduce the number of trapped atom as:

$$N = \frac{d^2}{\sigma}\left(\frac{\upsilon_c}{u}\right)^4 = \frac{4r^2\upsilon_R^2 d^4}{\sigma u^4}. \tag{3.23}$$

It can be seen that when the laser intensity is small, $N \propto d^4 I^2$. Increasing the laser intensity and beam diameter is very helpful for increasing the number of trapped atoms. When the laser intensity is high, $I \gg I_s$, the interaction between atoms and light has reached a saturated state, then the number of atoms cannot be increased any more by increasing the laser power.

In the process of loading atoms in the magneto-optical trap, the gradient of the magnetic field is also an important parameter. If it is too small, Γ_c will increase, resulting in a higher atomic escape rate, and if it is too large, the atomic loading rate R will increase. Therefore, in the process of optimizing the magneto-optical trap, this needs to be adjusted. The final gradient value used should also be optimized based on the number of atoms and their density.

3.5 Polarization Gradient Cooling

Since the magneto-optical trap is mainly based on the Doppler cooling technology, the temperature of the cold atoms obtained in the magneto-optical trap should not be lower than the Doppler limit for temperature, however, in real experiments, the temperature of the trapped cold atoms is lower than the Doppler temperature. This phenomenon was first discovered by the Phillips group in 1988 [13], and then Steven Chu's group also discovered a similar phenomenon [14]. Through further analysis, it was discovered that there is another kind of cooling mechanism in the process, which is called polarization gradient cooling.

In simple words, in the area where the six laser beams overlap, the laser field will be the superposition of the six beams. The polarization of the overlapped lasers will vary from position to position, i.e., there will be a polarization gradient between different positions. The potential energy of the interaction between atoms and lasers is related to the orientation of the electric dipole moment to the laser field, so the polarization difference of the laser field at different positions will inevitably lead to the difference of the interaction potential energy. When an atom at a low energy level moves from one place to another, if the potential energy increases, it will inevitably require a decrease in kinetic energy to compensate, and the increase in potential energy will make it easier for atoms to satisfy the conditions of resonant absorption and absorb photons. During spontaneous radiation, there is a certain probability of radiating to a lower energy level, resulting in a decrease in the net kinetic energy of the atoms during the whole process, so as to achieve cooling and deceleration of the atoms. Repeatedly, the velocity of the atoms is further reduced and atoms are cooled down.

Commonly used polarization gradient cooling methods have two combinations: one is the linear polarized lasers at opposite directions perpendicular to each other, called "lin⊥lin" configuration, and the other one is the circular polarized lasers at opposite directions perpendicular to each other, called "$\sigma^+ - \sigma^-$" configuration. In the experiment based on sodium atoms, in the case of "lin⊥lin", through the analysis of Dalibard et al. [15], the temperature to which the atom can be cooled can be approximately written as:

$$T \simeq -\frac{\hbar\Omega^2}{8k_b(\omega-\omega_0)}. \tag{3.24}$$

In the case of red detuning, this temperature is proportional to the laser power, and inversely proportional to the magnitude of the detuning. Therefore, reducing the-laser power and increasing the detuning is conducive to obtaining lower temperature atoms, but the power cannot be too small. When the power is too small, the lasers are impossible to enter the atom to participate in the atom-laser interaction or the interaction is very weak, which is obviously not feasible. For the "$\sigma^+ - \sigma^-$" configuration in cooling sodium atoms, the cooling temperature can also be calculated as:

$$T \simeq -\frac{\hbar\Omega^2}{k_b(\omega-\omega_0)}\left[\frac{29}{300}+\frac{254}{75}\frac{\Gamma^2/4}{(\omega-\omega_0)^2+\Gamma^2/4}\right]. \tag{3.25}$$

Similar with Eq. 3.24, the cooling effect is better with lower laser power and large detuning. This conclusion also suits to other alkali metal atoms. It can be seen that under the influence of polarization gradient cooling, atoms can be cooled to a temperature lower than the Doppler limit temperature. In our experiment, three pairs of lasers with perpendicular polarizations overlap, and the polarization of lasers in the intersection area varies with the position and it is adapted to the polarization gradient cooling, so that the temperature of the cold atomic ensemble lower than v_D can be finally obtained.

3.6 Cold Atomic Optical Depth and Temperature

After trapping the cold atoms in the magneto-optical trap, we need to measure the numbers, density, optical depth and temperature of the cold atoms. Usually, when measuring the atomic numbers, density and optical depth of the atomic ensemble, the absorption imaging method is used. When measuring the atomic temperature, it needs to be combined with time-of-flight measurement.

3.6.1 Measurement of Optical Depth via Absorption Imaging

After getting the cold atomic ensemble in the magneto-optical trap, the number of atoms in the atomic ensemble N, the size of the atomic ensemble (in this experiment, the atomic ensemble is approximately spherical Gaussian distribution with a radius of r) and the atomic number density n are very worthy of attention. In order to measure these parameters of atomic ensembles, several methods including fluorescence detection, absorption imaging detection, and non-destructive measurement can be used. Relatively speaking, the absorption imaging method is simple and easy to operate, so we use the absorption imaging method to measure these parameters.

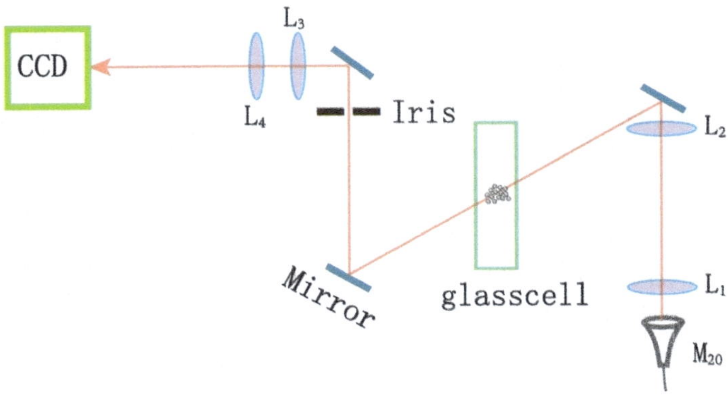

Fig. 3.13 Experimental setup for absorption imaging

As shown in Fig. 3.13, the imaging light in the optical fiber (D2 line $F = 2 \rightarrow F' = 3$ transition, Fig. 3.5) passes through a collimator of SK's M_{20}, after the collimating lens, it is approximately parallel light with a diameter of 4 mm, then the imaging light passes through two lenses L_1 (focal length $f_1 = 50$ mm) and L_2 (focal length $f_2 = 150$ mm) to enlarge the beam size 3 times, so that the imaging light can completely cover the atomic ensemble. After passing through the vacuum glasscell, the atoms will absorb the light, and the residual light enters the camera for detection. The iris is used to block the fluorescence from the atoms. The lens L_3 (focal length $f_3 = 300$ mm) and L_4 (focal length $f_4 = 250$ mm) are used for a size transformation from atoms to image. By observing the change of imaging light intensity before and after passing through the atoms, we can estimate the number of atoms. The measuring instrument is Andor's camera, which can identify the region of the light-absorbing atoms, thereby telling the size of the atomic ensemble.

In order to prevent the light from being easily saturated in the experiment, the imaging light we used usually has 3 MHz red detuning, and the incident light power is on the order of microwatts. In order to accurately obtain the atomic ensemble's information, we take a total of 3 images for each measurement, as show in Fig. 3.14. The first image: when the cold atom ensemble exists, we measure the light intensity distribution $I_1(\boldsymbol{r})$ after the image light passes through the atoms, as shown in the Fig. 3.14a; the second image: the same method as acquiring the first image under the condition of that the magnetic field is turned off so that the cold atomic ensemble doesn't exist. We also measure the light intensity distribution $I_2(\boldsymbol{r})$ when the light passes through the glasscell, as shown in Fig. 3.14b; the third image: we turn off the imaging light together without cold atoms, and measure the light intensity distribution $I_3(\boldsymbol{r})$ on the CCD caused by the background light, as shown in Fig. 3.14c. In this way, after subtracting the background light field, the final light intensity distribution of the imaging light with and without atoms can be obtained as $I_{out} = I_1(\boldsymbol{r}) - I_3(\boldsymbol{r})$, $I_{in} = I_2(\boldsymbol{r}) - I_3(\boldsymbol{r})$. The difference of light intensity before and after passing through the atoms $I_2(\boldsymbol{r}) - I_1(\boldsymbol{r})$ represents the absorption of light by the atoms, as shown in

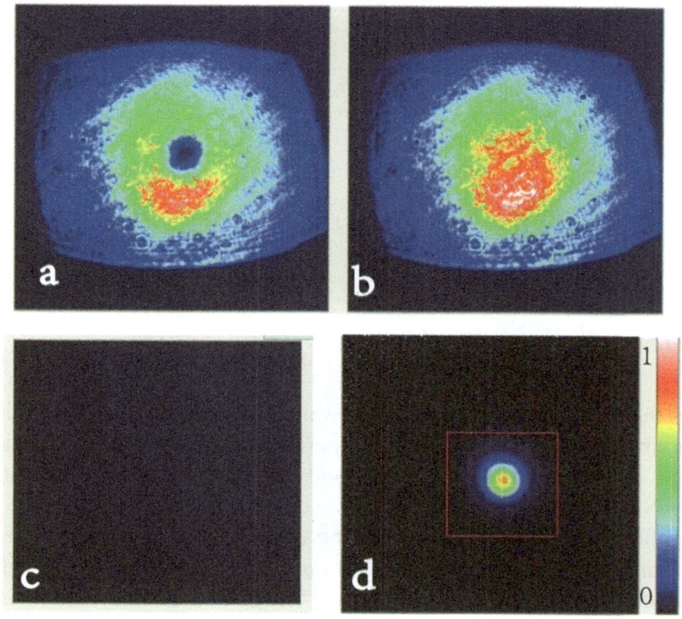

Fig. 3.14 Results for absorption imaging a the intensity distribution of the imaging light measured on the CCD after passing through the atoms. The dark spots are the areas where the atoms absorb light. **b** The intensity distribution of the imaging light on the CCD when the magnetic field of the magneto-optical trap is turned off and no cold atoms are trapped. **c** The background light distribution observed on the CCD when there is neither trapped atoms nor imaging light. **d** intensity distribution of atomic absorption (difference of Fig. **a** and **b**)

Fig. 3.14d. Suppose the light propagation direction is z, the atomic number density can be expressed as n, and the absorption cross-section of atoms is σ_{abs}. From the Lambert-Beer law, it can be known that when the imaging light does not reach the saturation intensity, the change of light intensity before and after the atom satisfies:

$$I_{out} = I_{in}e^{-\sigma_{abs}\int ndz} = I_{in}e^{-od}, \tag{3.26}$$

where $n_z = \int ndz = nd$ is the two-dimensional atomic number density of the atoms in the imaging light direction, d is the effective diameter of the atom, and od is the optical depth. Through measurements of absorption imaging, I_{out}, I_{in} and atomic size d can be obtained, from which the optical depth of the atoms can be calculated as:

$$od = \ln \frac{I_{in}}{I_{out}}. \tag{3.27}$$

The atomic number density[1] is:

$$n = \frac{od}{\sigma_{abs} d}. \tag{3.28}$$

Then the number of trapped atoms is:

$$N = n\frac{4\pi}{3}(d/2)^3 = \frac{\pi d^2 \cdot od}{6\sigma_{abs}}. \tag{3.29}$$

3.6.2 Measurement of Cold Atomic Temperature

Because in the follow-up experiments, we want the atoms to stay in a specific area as long as possible, and not move out of the experimental area too quickly due to the high temperature, it requires the measurement and optimization of the temperature of the cold atoms. Since the temperature of atoms is closely related to its velocity of motion, measuring the temperature of atoms can be equivalently changed to measuring its velocity of motion. If the magneto-optical trap is turned off, the previously trapped atoms will move freely and start to expand. The size of atomic ensemble becomes bigger and bigger, and the speed of this expansion is directly related to the velocity (or temperature) of the atoms. Therefore, by measuring the size of the atomic ensemble expansion at different times, the velocity can be measured, and the temperature of the atomic ensemble can be further obtained. This method is called the time-of-flight measurement.

Taking a one-dimensional case as an example, the internal velocity distribution of the atomic ensemble trapped in the magneto-optical trap with a temperature of T is:

$$f(\upsilon) = \sqrt{\frac{m}{2\pi k_b T}} Exp\left[-\frac{m\upsilon^2}{2k_b T}\right]. \tag{3.30}$$

When the magnetic field and light are turned off, atoms will move freely. Replace υ in the above formula with $\upsilon = \frac{x - x_0}{t}$, where x represents the position of the atom in one-dimensional case, so we rewrite it as:

$$f(x - x_0, t) = \sqrt{\frac{m}{2\pi k_b T}} Exp\left[-\frac{m(x - x_0)^2/t^2}{2k_b T}\right]. \tag{3.31}$$

$$n_0(x) = \sqrt{\frac{1}{2\pi r_0^2}} Exp\left(-\frac{x^2}{2r_0^2}\right). \tag{3.32}$$

[1] In fact, the distribution of the number of atoms is approximately Gaussian, with dense in the center and sparse near the edges. Here is just an estimate of its density.

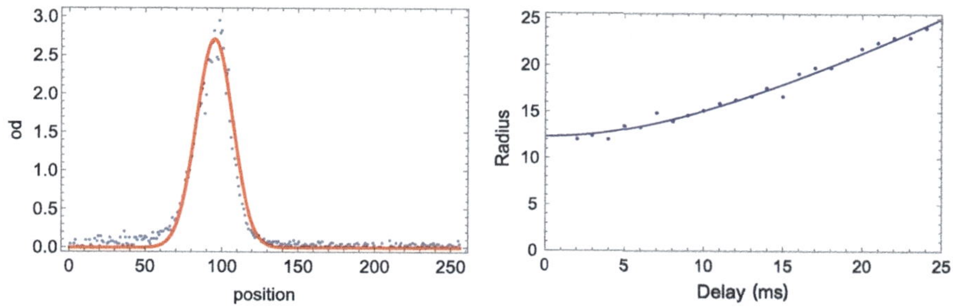

Fig. 3.15 Results for measurements of *od* and temperature of cold atoms a the density distribution of atoms is approximately Gaussian, with the largest number of central atoms, the value of *od* discussed in the following experiments is the central maximum value obtained according to this fitting. **b** The change of the radius of the atomic ensemble with the expansion time after the atoms freely move (Eq. 3.35). The real size of the position and radius here has not been converted according to the pixel and imaging size

As the density distribution of atoms is also considered as the Gaussian distribution, i.e.,

$$n_0(x) = \sqrt{\frac{1}{2\pi r_0^2}} Exp\left(-\frac{x^2}{2r_0^2}\right), \tag{3.33}$$

where r_0 is atomic ensemble's radius. After time t, the atomic ensemble becomes:

$$n(x,t) = \int dx_0 f(x - x_0, t) n_0(x_0) \propto Exp\left(-\frac{x^2}{2r^2}\right). \tag{3.34}$$

The radius becomes:

$$r^2 = r_0^2 + k_b T t^2/m \tag{3.35}$$

Now the temperature T of the atomic ensemble can be obtained by measuring the radius of the atomic ensemble $r(t)$ at different diffusion moments, and then fitting radius-time curve, as shown in Fig. 3.15.

In our experiment, the typical experimental parameter results of the cold atoms trapped in the magneto-optical trap are:
Radius of cold atomic ensemble r_0: 360 μm.
Optical depth *od*: 4.5.[2]
Number of atoms N: 1.4×10^7.
Atomic number density n: $7 \times 10^{10}\,\text{cm}^{-3}$.
Atomic temperature T: 82 μK.

If the loading time of the magneto-optical trap is increased or the number of background atoms in the vacuum chamber is increased, the number of cold atoms that can be trapped can be further increased.

[2] For light with different wavelengths of 780 and 795 nm, *od* is different due to the transition coefficient, usually the former is twice the latter on average.

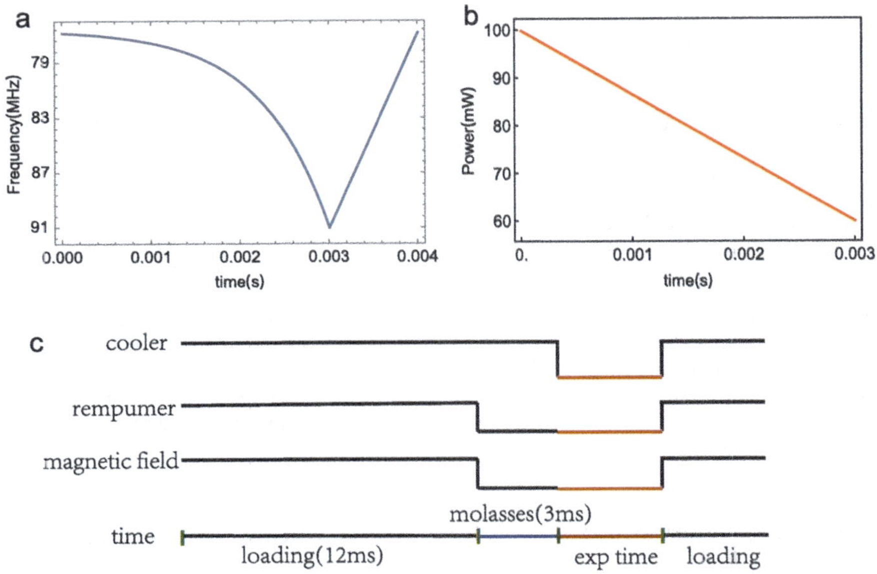

Fig. 3.16 Optical molasses process of dynamically adjusting frequency and power of cooler laser. a Dynamically adjust the detuning of cooler laser to further cool the atoms. The last 1 ms is to restore the built-in frequency of the radio frequency source to prepare for the next round of loading atoms to prevent unnecessary trouble caused by frequency sudden change in next round of molasses, during the 1 ms, the output of the radio frequency source is closed. **b** The power change of cooler laser during the molasses process. **c** Time sequences for MOT loading

3.7 Further Cooling of Cold Atoms

From the results of the previous section, we can see that the temperature of the atoms directly trapped by the magneto-optical trap is in the order of hundreds of micro-Kelvin. In order to further lower the atomic temperature, we have made further operations on the polarization gradient cooling of the atoms (that is, optical molasses). As described in Eqs. 3.24 and 3.25, the cooling temperature of atoms is proportional to the laser power, and inversely proportional to the detuning. Therefore, in order to obtain a lower temperature, we dynamically adjust intensity and detuning of the cooler laser.

As shown in Fig. 3.16a is the varying curve for the frequency and intensity of cooler laser when the atoms are further cooled in the experiment. Within 3 ms after the magnetic field is turned off, we dynamically adjust the detuning from −20 to −50 MHz, scaling as $e^{-\frac{t}{0.75\,\text{ms}}}$, and the laser intensity from 100 to 60 mW.

When only molasses with varying intensity is made, it can be clearly observed that the temperature of the atoms is reduced from 100 μK to 30 μK. If the frequency sweeping molasses are further applied, the temperature can be further reduced to around 10 μK. During the entire molasses process, although the temperature of the atoms becomes lower, the change for number and density of atoms is not particularly obvious, they remain at the same order of magnitude as before.

3.8 Summary

In this chapter, I introduce the fundamentals and experimental settings for the cold atomic ensemble, including laser cooling, measurement methods for cold atoms, and the settings of a series of experimental conditions such as vacuum, laser, and optical path, etc. In the end, we can prepare cold atomic ensemble with $od = 5$ and the temperature of 10 μK, the prepared cold atoms provide favorable prerequisites for the following experiments.

References

1. Mo J-Y (2009) A Laser Diode system stabilized on the saturated absorption lines of potassium atoms:[Bachelor Thesis]. University of Science and Technology of China
2. Chen X-Z (2005) Principle and technology of frequency stabilization of semiconductor lasers
3. Andrews JR (2000) Broadband coaxial bias tees. Picosecond Pulse Labs
4. Drever RWP, Hall JL, Kowalski FV, Hough J, Ford GM, Munley AJ, Ward H (1983) Laser phase and frequency stabilization using an optical resonator. Appl Phys B 31(2):97–105
5. Jiang X (2009) The Laser frequency stabilizing and phase locking on quantum memory based on cold atoms. PhD thesis, University of Science and Technology of China
6. Hänsch TW, Schawlow AL (1975) Cooling of gases by laser radiation. Opt Commun 13(1):68–69
7. Migdall AL, Prodan JV, Phillips WD, Bergeman TH, Metcalf HJ (1985) First observation of magnetically trapped neutral atoms. Phys Rev Lett 54(24):2596
8. Chu S, Hollberg L, Bjorkholm JE, Cable A, Ashkin A (1985) Three-dimensional viscous confinement and cooling of atoms by resonance radiation pressure. Phys Rev Lett 55(1):48
9. Foot CJ et al (2005) Atomic physics, vol 7. Oxford University Press
10. Ketterle W, Davis KB, Joffe MA, Martin A, Pritchard DE (1993) High densities of cold atoms in a dark spontaneous-force optical trap. Phys Rev Lett 70(15):2253
11. Steck DA (2001) Rubidium 87 d line data
12. Gibble KE, Kasapi S, Chu S (1992) Improved magneto-optic trapping in a vapor cell. Opt Lett 17(7):526–528
13. Lett PD, Watts RN, Westbrook CI, Phillips WD, Gould PL, Metcalf HJ (1988) Observation of atoms laser cooled below the doppler limit. Phys Rev Lett 61(2):169
14. Shevy Y, Weiss DS, Ungar PJ, Chu S (1989) Bimodal speed distributions in laser-cooled atoms. Phys Rev Lett 62(10):1118
15. Dalibard J, Cohen-Tannoudji C (1989) Laser cooling below the doppler limit by polarization gradients: simple theoretical models. JOSA B 6(11):2023–2045

Chapter 4
Highly Retrievable Quantum Memories

In quantum networks, the basic building block is to realize quantum connection of separate quantum nodes. If one wants to efficiently realize the connection between quantum memories with cold atomic ensembles, the performance of a single quantum memory is required to be high enough to be able to obtain high-quality photon-atom interface. One of the most important parameters to assess the performance of a quantum memory is retrieval efficiency, which indicates how efficiently the stored quantum information can be retrieved. Therefore, in this chapter, I will mainly introduce the experimental progress of DLCZ quantum memory with cold atomic ensembles, analyze and optimize the retrieval efficiency affected by the atomic optical depth, spin wave motion and other factors, so as to improve the retrieval efficiency of quantum memory.

4.1 Background

Quantum memories [1–3] play an essential role in quantum communication, quantum computing, quantum precision measurement and other fields [4–8], so it is of great significance to develop high-performance quantum memories. A key element in quantum memory is the storage and retrieval efficiency. Taking the quantum repeater scheme [4] as an example, an efficiency increase of 1% can reduce the time required for entanglement distribution by 7–18% [9], the realization of highly efficient and retrievable quantum memory has become a research field for many research groups. Among the various quantum memory protocols, so far, 87% efficiency has been achieved in GEM storage [10], and 82% efficiency of Raman protocol in cold atomic ensemble has been achieved [11]. Based on EIT protocol with cold atomic ensembles, the storage and retrieval efficiency of 68% [12] for single photons, and 92% [9] for weak coherent light has been reported. In our experiments, the goal is to reach a quantum communication distance over 1,000 km based on the quantum repeater

B. Jing, *Quantum Network with Multiple Cold Atomic Ensembles*, Springer Theses,
https://doi.org/10.1007/978-981-19-0328-1_4

scheme [13] with DLCZ quantum memories. The most basic block is the realization of DLCZ quantum memory based on the cold atomic ensembles. This protocol is similar with electromagnetically induced transparency (EIT) protocol [14, 15], while the photon-atom interface can be directly generated through spontaneous Raman scattering in advance (Fig. 1.6). The whole process of DLCZ quantum memory and its quantization are introduced in the following sections.

4.2 DLCZ Quantum Memory and Its Quantization

In quantum memories with DLCZ protocol, the main processes are the 'write' process and 'read' process, as mentioned in Sect. 1.4, here I will detail the write and read process.

4.2.1 *The Write Process*

As shown in Fig. 1.6, all atoms are initially prepared in a ground state g, then a write beam with a detuning of Δ_w from excited states e interacts with the atoms, the write beam is described as:

$$\boldsymbol{E}_w(z,t) = \hat{\epsilon}_w E_w(z,t)e^{i(k_w z - \omega_w t)} + h.c. \tag{4.1}$$

Through spontaneous Raman scattering, one atom will transfer to s with a very low probability and emit a stokes photon in the direction of $\boldsymbol{r}$ of the entire 4π three-dimensional space (write-out photon), the field operator of this photon can be described as [16]:

$$\boldsymbol{E}_{wo}(\boldsymbol{r},t) = \sum_k \hat{\epsilon}_k \varepsilon_k a_k e^{i(k_{wo}\cdot \boldsymbol{r} - \omega_{wo} t)} + h.c., \tag{4.2}$$

where $\varepsilon_k = \sqrt{\frac{\hbar\omega_k}{2\epsilon_0 V}}$, ω_k is the frequency of write beam, a_k is the annihilation operator. As cold atomic temperature is around 10 μK, the atom velocity is about 3 cm/s. During the write process, the pulse width of write beam is on the order of several tens of nano-seconds, so the atomic moving distance can be ignored. The ith atomic position is denoted as $\boldsymbol{r}_i$. The Hamiltonian of photon-atom interaction during write process can be written as:

$$H = \sum_{i=1}^{N}\{\hbar\Delta_w\sigma_{ee}^i + [-\hbar\Omega_w(\boldsymbol{r}_i,t)e^{ik_w\boldsymbol{r}_i}\sigma_{eg}^i + \sum_k \hbar g_k a_k e^{i(\boldsymbol{k}\boldsymbol{r}_i - \Delta\omega_k t)}\sigma_{es}^i + h.c.]\}. \tag{4.3}$$

Here $\sigma_{xy} = |x\rangle\langle y|(x, y = g, e, s)$, $\Delta\omega_k = \omega_k - \omega_w - \omega_{sg}$, $g_k = -\frac{d_{es}\hat{\epsilon}_k\epsilon_k}{\hbar}$ is the coupling coefficients of atoms and write-out photons, $\Omega_w = \frac{d_{eg}\hat{\epsilon}_w E_W}{\hbar}$ is the Rabi frequency of the write beam.

If the weak write beam keeps detuned from the excited state, the excited state in the above equation can be adiabatically eliminated, then the whole write process can be simplified to only two states g and s involved. Under the adiabatic approximation and ignoring the atom-atom interaction, Eq. 4.3 can be rewritten as [17, 18]:

$$H = \sum_{i=1}^{N}\left[\sigma_{sg}^{i}\frac{\Omega_w(\boldsymbol{r}_i, t)e^{ik_w\boldsymbol{r}_i}}{\Delta_w}\sum_k \hbar g_k \hat{a}_k^{\dagger} e^{-i(k\boldsymbol{r}_i - \Delta\omega_k t)} + h.c.\right]. \tag{4.4}$$

In the adiabatic process, it can be assumed that there is a virtual excited energy level whose line width Γ' and the line width Γ of the real energy level e satisfy $\Gamma' = \frac{\Omega_w^2}{\Delta_w^2}$, the width of the write light pulse T is much smaller than $1/\Gamma'$. The whole write process for generating write-out photons can be regarded as a spontaneous Raman process rather than a stimulated radiation process. By solving the first-order perturbation of the Schrodinger equation, the photonic and atomic states in the write process can be described as:

$$\begin{aligned} |\Psi\rangle &= \left[1 - i\int_0^T H(t)dt\right]|vac\rangle \\ &= |vac\rangle + \sum_{i=1}^{N}\sigma_{sg}^{i}\frac{\Omega_w(\boldsymbol{r}_i, t)e^{ik_w\boldsymbol{r}_i}}{\Delta_w}|g, \ldots s_i, \ldots, g\rangle|1\rangle_{wo,i}. \end{aligned} \tag{4.5}$$

Here the vacuum state $|vac\rangle = |g\rangle_a^{\otimes N}|0\rangle_{wo}$ means the initial state, i.e., all atoms are in state g and without write-out photons. $|1\rangle_{wo,i} = -i\int_0^T dt\sum_k \hbar g_k a_k^{\dagger} e^{-i(\boldsymbol{k}\cdot\boldsymbol{r}_i - \Delta\omega_k t)}|0\rangle_{wo}$ represents the joint atom and photon state when a write-out photon is emitted via the ith atom. For the atomic ensemble, from Eq. 2.5, it can be seen that the light field scattered by the ith atom (position at $\boldsymbol{r}_i$) is proportional to $\frac{e^{ik_{wo}(\boldsymbol{r}-\boldsymbol{r}_i)}}{(\boldsymbol{r}-\boldsymbol{r}_i)}$. In our experiments, the angle between write beam and write-out photons is small enough ($\sim 3°$), with paraxial approximation, we can deduce [19]:

$$E_i(\boldsymbol{r}) \propto e^{-i\boldsymbol{k}_{wo}\cdot\boldsymbol{r}_i}. \tag{4.6}$$

Going further, considering only the write process of a specific transition in a single frequency mode, the state with one write-out photon can be written as

$$|1\rangle_{wo,i} \propto a_{wo}^{\dagger}e^{-i\boldsymbol{k}_{wo}\cdot\boldsymbol{r}_i}|0\rangle_{wo}. \tag{4.7}$$

Combine with Eq. 4.5, the joint atom and photon state after emitting a write-out photon can be described as:

$$|\Psi\rangle = (1 + \sqrt{\chi}\sum_i e^{i\Delta\boldsymbol{k}\cdot\boldsymbol{r}_i}\sigma_{sg}^i a_{wo}^\dagger)|vac\rangle, \tag{4.8}$$

where χ refers to the probability that a single atom can be excited, named as the excitation probability, which is a parameter commonly used in experiments that depends on the intensity of the interaction between light and atoms. It is related to the write beam power, detuning, atomic transitions, and solid angle for collecting photons. $\Delta\boldsymbol{k} = \boldsymbol{k}_w - \boldsymbol{k}_{wo}$ is the difference between the wave vector of the write-out photon and the write beam. After the selection of a write-out photon with specific direction and frequency mode, we can see that the phase is related to the position of the atom. By defining the collective raising operator of atomic ensemble:

$$S^\dagger = \frac{1}{\sqrt{N}}\sum_i e^{i\Delta\boldsymbol{k}\cdot\boldsymbol{r}_i}\sigma_{sg}^i, \tag{4.9}$$

Equation 4.8 can be simplified as:

$$|\Psi\rangle = (1 + \sqrt{p}S^\dagger a_{wo}^\dagger)|vac\rangle. \tag{4.10}$$

Here $p = N\chi$ is the excitation probability of emitting a write-out photon from the atomic ensemble. It can be seen that under the condition of weak write beam, $p \ll 1$, there is a high probability that no write-out photons are generated, all atoms are still in the initial states, however, the successful case of write process can always be selected by detecting the write-out photons. The main reason why the excitation probability is set to be very low in the experiment is that when the excitation probability is high, the write beam will excite two atoms with a probability of p^2, as a result, multiple excitations will affect the fidelity of quantum memory. In order to ensure that the write-out photons are mainly single photons, it is necessary to use a low excitation probability, so that when a write-out photon is detected, the probability of two or more excitations in the atomic ensemble will be very low. This situation can be completely avoided by operate the write process based on the Rydberg blockade mechanism to generate a deterministic single photon [20], where if one atom is excited to Rydberg sate, there will be no multiple excitations.

4.2.2 The Read Process

From Eq. 4.8, it is known that if a write-out photon is collected in a specific solid angle and detected, the atomic states in the atomic ensemble will be in a collective state:

$$|\Psi\rangle_a = S^\dagger|vac\rangle_a. \tag{4.11}$$

After some time, when the atoms move to another position $\boldsymbol{r}_i'$, a strong read beam is applied to the atoms, similar with the write process, here we define a collective lowering operator for atomic ensemble as:

$$S = \sum_i e^{i(\boldsymbol{k_r} - \boldsymbol{k_{ro}})\cdot \boldsymbol{r}'_i} \sigma_{gs}^i. \tag{4.12}$$

Then the joint state between atoms and photons during the read process can be written as:

$$\begin{aligned}|\Psi\rangle_f &= S a_{ro}^{\dagger} |\Psi\rangle_a |0\rangle_{ro} \\ &= |vac\rangle_a \otimes \frac{1}{\sqrt{N}} \sum_{i=1}^{N} e^{i[(\boldsymbol{k_w} - \boldsymbol{k_{wo}})\cdot \boldsymbol{r}_i + (\boldsymbol{k_r} - \boldsymbol{k_{ro}})\cdot \boldsymbol{r}'_i]} a_{ro}^{\dagger} |0\rangle_{ro},\end{aligned} \tag{4.13}$$

from which the read-out photonic state can be extracted as:

$$|\Psi\rangle_{ro} = \frac{1}{\sqrt{N}} \sum_{i=1}^{N} e^{i[(\boldsymbol{k_w} - \boldsymbol{k_{wo}})\cdot \boldsymbol{r}_i + (\boldsymbol{k_r} - \boldsymbol{k_{ro}})\cdot \boldsymbol{r}'_i]} a_{ro}^{\dagger} |0\rangle_{ro}. \tag{4.14}$$

As mentioned in Chap. 2, if the system is a single atom instead of an atomic ensemble and the probability that the read-out photon will exit along the wave vector $\boldsymbol{k_{ro}}$ is p_{ro}, then this probability for the atomic ensemble will become:

$$P_{ro} = \frac{1}{N} \left| \sum_{i=1}^{N} e^{i[(\boldsymbol{k_w} - \boldsymbol{k_{wo}})\cdot \boldsymbol{r}_i + (\boldsymbol{k_r} - \boldsymbol{k_{ro}})\cdot \boldsymbol{r}'_i]} \right|^2 p_{ro}. \tag{4.15}$$

If we do not consider the storage time and ignore the movement of the atoms, i.e., $\boldsymbol{r}_i = \boldsymbol{r}'_i$, when the phase-matching condition is met:

$$\boldsymbol{k_w} - \boldsymbol{k_{wo}} + \boldsymbol{k_r} - \boldsymbol{k_{ro}} = 0. \tag{4.16}$$

Under this condition, $P_{ro} = N p_{ro}$, in contrast with the case for a single atom, the probability of collecting read-out photons in the phase-matching direction will be increased by N times, this enhancement effect is the collective enhancement effect. Next, we consider the calculation of the retrieval efficiency under the collective enhancement effect in the real situation [21].

Still considering the read process, assuming the probability of emitting read-out photons per unit solid angle along any direction is p_{ro}. Due to the collective enhancement effect, the probability of emitting read-out photons along the direction under the specified phase-matching conditions in the experiment will increase N times, and the solid angle range of the read-out photons that can be collected under the phase-matching condition should be the solid divergence angle of the read-out photon at the center of the atom (The beam waist should be approximate the radius of the central atomic ensemble that can participate in the write process and emit write-out photons that are collected.), that is, $\Delta\Omega = \pi(\frac{\lambda}{\pi w_0})^2$, then we consider that the atoms fall to the specific energy level due to competition at different transitions during the read process, there is a probability ξ_g (that is, the square of the dipole transition coefficient) with which the atoms fall into the initial energy level. The probability that the photon can be effectively read out in the phase-matching direction during the entire read process can be obtained as:

$$P_{\Delta\Omega} = p_{ro}\xi_g N\Delta\Omega = p_{ro}\xi_g N\pi(\frac{\lambda}{\pi w_0})^2, \tag{4.17}$$

while the probability of emitting the photons in other directions is:

$$P_{4\pi} = (4\pi - \Delta\Omega)p_{ro} \sim 4\pi p_{ro}. \tag{4.18}$$

The retrieval efficiency is then calculated as:

$$\eta_r = \frac{P_{\Delta\Omega}}{P_{\Delta\Omega} + P_{4\pi}} = \frac{\frac{N\xi_g\lambda^2}{4\pi^2 w_0^2}}{\frac{N\xi_g\lambda^2}{4\pi^2 w_0^2} + 1} = \frac{C}{C+1}. \tag{4.19}$$

Here,

$$C = \frac{N\xi_g\lambda^2}{4\pi^2 w_0^2} = \frac{\xi_g\lambda^2}{4\pi}\frac{N}{\pi w_0^2} \sim \frac{\xi_g\lambda^2}{4\pi}nL = \frac{\xi_g\lambda^2}{4\pi}\frac{od}{\sigma}, \tag{4.20}$$

σ is the absorption cross section of rubidium atoms, n is the atomic number density, L is the longitudinal size of the atomic ensemble, and od is the optical depth of the atomic ensemble.

When a read-out photon is generated anywhere in the atomic ensemble, it will constitute the electromagnetically induced transparency (EIT) process together with the strong read beam [14]. It is precisely because of the existence of EIT, the generated read-out photons will not be re-absorbed by the large number of atoms in the ground state g and thus can be measured. Instead, they will pass through the atomic ensemble through the EIT process and be collected.

4.2.3 EIT in the Read Process

Now we consider the efficiency of the read-out photon that can pass through the atomic medium in the EIT process. It will also directly affect the final retrieval efficiency. For simplicity and following the naming rules in EIT, the read-out photon is named as the probe beam (frequency is ω_p, the Rabi frequency is $\Omega_p = \boldsymbol{d_{eg}} \cdot \boldsymbol{E_p}$), the strong read beam is named as control beam (frequency is ω_c, the Rabi frequency is Ω_c, $\Omega_c \gg \Omega_p$). As shown in Fig. 4.1, the probe beam is detuned from excited state e with a detuning of Δ_p, the control beam is detuned from excited state e with a detuning of Δ_c. Let define two-photon detuning as $\delta = \Delta_p - \Delta_c$ and assume the

Fig. 4.1 EIT process in Λ-structure atoms

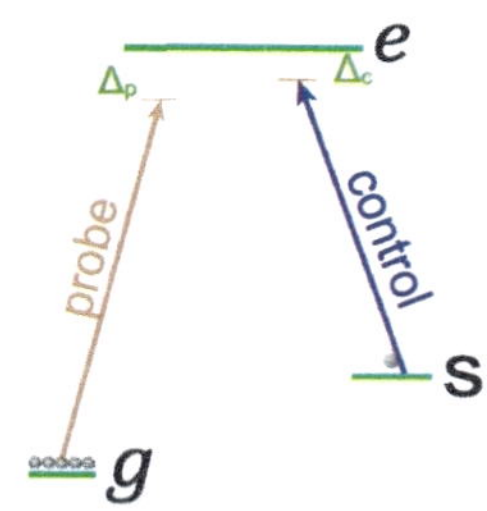

two beams are collinear where both are along the z direction. The Hamiltonian in the basis of g, s, and e can be written as:

$$H = -\frac{1}{2}\begin{pmatrix} 0 & 0 & \Omega_p \\ 0 & 2\delta & \Omega_c \\ \Omega_p & \Omega_c & 2\Delta_p \end{pmatrix}. \tag{4.21}$$

Through Lindblad equation [14, 22], the density matrix for the atomic sates should satisfy:

$$\begin{aligned} \frac{d\rho}{dt} =& \frac{1}{i}[H, \rho] + \frac{\Gamma_{eg}}{2}(2\sigma_{ge}\rho\sigma_{eg} - \sigma_{ee}\rho - \rho\sigma_{ee}) \\ &+ \frac{\Gamma_{es}}{2}(2\sigma_{se}\rho\sigma_{es} - \sigma_{ee}\rho - \rho\sigma_{ee}) + \frac{\Gamma_{sg}}{2}(2\sigma_{gs}\rho\sigma_{sg} - \sigma_{ss}\rho - \rho\sigma_{ss}) \end{aligned}. \tag{4.22}$$

In ideal cases, the dipole transition between the two ground states g and s is forbidden, i.e., Γ_{sg} can be considered as zero, however, due to various mechanisms such as atomic collisions, atoms in state s still has a certain chance to return to g, where it still satisfies the condition $\Gamma_{sg} \ll \Gamma_e = \Gamma_{eg} + \Gamma_{es}$. Equation 4.22 can be expanded as:

$$\begin{aligned} \dot{\rho}_{eg} &= \left(-\frac{\Gamma_e}{2} + i\Delta_p\right)\rho_{eg} + \frac{i}{2}\Omega_p(\rho_{gg} - \rho_{ee}) + \frac{i}{2}\Omega_c\rho_{sg} \\ \dot{\rho}_{es} &= \left(-\frac{\Gamma_e}{2} + i\Delta_c\right)\rho_{es} + \frac{i}{2}\Omega_p\rho_{gs} + \frac{i}{2}\Omega_c(\rho_{ss} - \rho_{ee}) \\ \dot{\rho}_{sg} &= \left(-\frac{\Gamma_{sg}}{2} + i\delta\right)\rho_{sg} - \frac{i}{2}\Omega_p\rho_{se} + \frac{i}{2}\Omega_c\rho_{eg} \end{aligned}. \tag{4.23}$$

During the experiment, as the intensity of probe beam is small enough, most of the atoms are still in state g, i.e., $\rho_{gg} - \rho_{ee} \simeq 1$. Also considering the intensity of control beam is larger than that of probe beam, i.e., $\Omega_c \gg \Omega_p$, the equations above can be simplified as:

$$\begin{aligned}\dot{\rho}_{eg} &= \left(-\frac{\Gamma_e}{2}+i\Delta_p\right)\rho_{eg}+\frac{i}{2}\Omega_p+\frac{i}{2}\Omega_c\rho_{sg}\\ \dot{\rho}_{es} &= \left(-\frac{\Gamma_e}{2}+i\Delta_c\right)\rho_{es}+\frac{i}{2}\Omega_p\rho_{gs}+\frac{i}{2}\Omega_c(\rho_{ss}-\rho_{ee}).\\ \dot{\rho}_{sg} &= \left(-\frac{\Gamma_{sg}}{2}+i\delta\right)\rho_{sg}+\frac{i}{2}\Omega_c\rho_{eg}\end{aligned} \tag{4.24}$$

From the Maxwell-Bloch equation, the dynamic evolution of the probe beam satisfies [9]:

$$\frac{\partial\Omega_p}{\partial z}+\frac{1}{c}\frac{\partial\Omega_p}{\partial t}=i\frac{d\Gamma_e}{2L}\rho_{eg}, \tag{4.25}$$

where d is the optical depth, L is the length of atomic medium. Take the Fourier transform of ρ_{eg}, ρ_{sg}, and Ω_p $(t\to\omega)$ to get F_{eg}, F_{sg}, F_p, and substitute them into the above equations, we can get:

$$\begin{aligned}-i\omega F_{eg} &= \left(-\frac{\Gamma_e}{2}+i\Delta_p\right)F_{eg}+\frac{i}{2}F_p+\frac{i}{2}\Omega_c F_{sg}\\ -i\omega F_{sg} &= \left(-\frac{\Gamma_{sg}}{2}+i\delta\right)F_{sg}+\frac{i}{2}\Omega_c F_{eg}\\ \frac{\partial F_p}{\partial z}-\frac{i\omega}{c}F_p &= i\frac{d\Gamma_e}{2L}F_{eg}\end{aligned} \quad . \tag{4.26}$$

By solving the equations,

$$F_{eg}=\frac{-[i(\delta+\omega)-\Gamma_{sg}/2]iF_p/2}{[i(\omega+\Delta_p)-\Gamma_e/2][i(\delta+\omega)-\Gamma_{sg}/2]+\Omega_c^2/4}. \tag{4.27}$$

Thereby, the evolution of the probe beam in the atomic medium can be obtained:

$$F_p(\omega,z)=F_p(\omega,0)Exp\{\frac{i\omega z}{c}+\frac{d\Gamma_e z}{4L}\frac{i(\delta+\omega)-\Gamma_{sg}/2}{[i(\omega+\Delta_p)-\Gamma_e/2][i(\delta+\omega)-\Gamma_{sg}/2]+\Omega_c^2/4}\} \quad . \tag{4.28}$$

For any photonic wavepacket's waveform $\Omega_p(t,z)$, the equation may not have an analytical solution, but it is always possible to use numerical methods to obtain the residual components of the light after passing through the atomic medium via inverse Fourier transform $\Omega_p'(t,z=L)$. In particular, if $\omega=0$, $z=L$, the transmission spectrum of the probe beam passing through the atomic medium can be expressed as:

$$\begin{aligned}T(\Delta_p) &= Re\left[\frac{F_p(0,L)}{F_p(0,0)}\right]^2\\ &= Re\left\{Exp\left[\frac{d\Gamma_e}{4}\frac{i\delta-\Gamma_{sg}/2}{(i\Delta_p-\Gamma_e/2)(i\delta-\Gamma_{sg}/2)+\Omega_c^2/4}\right]\right\}^2.\end{aligned} \tag{4.29}$$

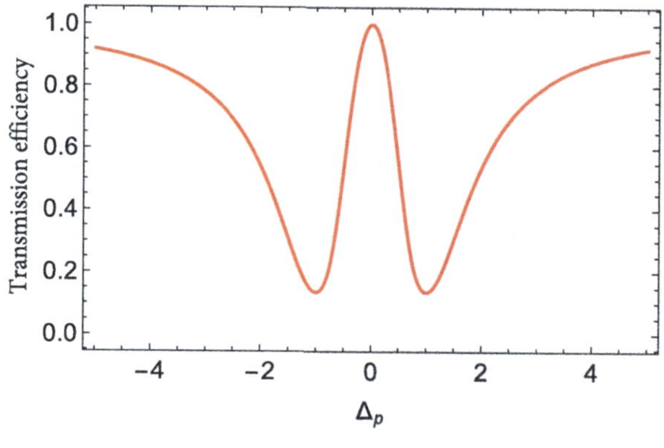

Fig. 4.2 Transmission spectrum of probe beam in EIT process

In an ideal situation, $\Gamma_{sg} = 0$, the maximum transmission efficiency can always reach the maximum value of 100%, and there will be no photonic loss during the transmission process. If the detuning of control beam detuning is set to zero, i.e., $\Delta_c = 0$, it can be seen that in the EIT process, the transmission spectrum of the probe beam changes with its detuning Δ_p, as shown in Fig. 4.2, when $\Delta_p = 0$, two-photon resonance is satisfied, the intensity of transmitted light is the highest, atomic ensemble doesn't absorb any probe light. The atomic ensemble is then equivalent to a transparent medium to the probe beam. The process is also the origin of electromagnetically induced transparency. The detuning width when the transmission efficiency drops to 1/2 near the resonance line is called as the transmission window $\Delta\omega_{EIT}$, as can be calculated from Eq. 4.29:

$$\Delta\omega_{EIT} \propto \frac{\Omega_c^2}{\sqrt{d}}. \tag{4.30}$$

Now it is clear that the transmission window of EIT can be controlled by the intensity of control beam and optical depth of atomic medium. As Γ_{sg} is not zero in real experiment, the probe beam will be inevitably absorbed while passing through the atoms, the absorption is related to intensity of control beam and atomic optical depth.

If we only care about the absorption and dispersion of the probe beam before and after passing through the atom rather than the evolution of $\Omega_p(t, z)$ in the whole process, we can directly solve the susceptibility in the EIT process in Eq. 4.24, the result is [14]:

$$\begin{aligned} \chi =& -\frac{|\boldsymbol{d}_{ge}|^2 n}{\epsilon_0} \frac{\delta + i\Gamma_{sg}/2}{(i\Delta_p - \Gamma_e/2)(i\delta - \Gamma_{sg}/2) + \Omega_c^2/4} \\ =& \chi' + i\chi'', \end{aligned} \tag{4.31}$$

where the imaginary part χ'' corresponds to the absorption of light before entering and after leaving the atoms, while the real part χ' corresponds to the dispersion of light before entering and after leaving the atoms. n is the atomic numbers density. The transmission efficiency of the probe beam after passing through atoms can also be obtained:

$$\eta_t = Exp(-kL\chi''). \tag{4.32}$$

Considering the DLCZ quantum memory in our experiment, we can finally obtain the read-out photon efficiency from the atomic ensemble as:

$$\eta = \eta_r \eta_t. \tag{4.33}$$

This efficiency not only depends on the phase-matching condition and collective enhancement effect in the read-out process, which makes the read-out photons emit with a high probability in the specific direction, but also is limited by the absorption of the emitted read-out photons during the EIT process that exists in the retrieval process. In our experiments, the former can be solved by increasing the atomic optical depth to improve the collective enhancement effect, and the followed EIT process can be controlled by increasing the light intensity and reducing the atomic optical depth. Therefore, choosing the appropriate optical depth and the appropriate intensity of read beam are important to realize a highly retrievable DLCZ quantum memory.

4.2.4 Performance Criteria for Quantum Memory

For a given quantum memory, everyone will have questions about whether it is good or bad. How efficiently can the stored information be retrieved? Is the retrieved quantum information same with the information stored? How long can the information be stored in the memory? What is the memory capacity and storage bandwidth? This series of issues are not only attracting attention in classical memory, but also very important in quantum memory. For the DLCZ quantum memory we are exploring, the retrieval efficiency, storage lifetime, fidelity, and the correlation function between write-out photons and read-out photons are important parameters that we use to evaluate the quantum memory. Since this thesis mainly focuses on the realization of highly efficient and retrievable quantum memory, the most important thing in the experiment is the retrieval efficiency. In the previous section, the read-out process and the influencing factors of the retrieval efficiency have been explained. Here, I will briefly introduce the remaining parameters used in the experiment.

4.2.5 Storage Lifetime

In DLCZ quantum memory, the quantum state is stored in the atoms, as the atomic states will decohere because of either interacting with the environments or its own movement, the stored quantum state will not be maintained forever, and the retrieval efficiency will also decrease with the increasing of storage time. Similar to the classic memory, we define the storage lifetime as the storage time when the retrieval efficiency drops to $1/e$ of the initial efficiency. In Eq. 4.14, we can see that the retrieval efficiency is related to the phase φ_i accumulated in the storage process:

$$\begin{aligned}\varphi_i &= (\boldsymbol{k}_w - \boldsymbol{k}_{wo}) \cdot \boldsymbol{r}_i + (\boldsymbol{k}_r - \boldsymbol{k}_{ro}) \cdot \boldsymbol{r}_i' \\ &= \Delta \boldsymbol{k} \cdot (\boldsymbol{r}_i' - \boldsymbol{r}_i).\end{aligned} \tag{4.34}$$

As atoms are moving with a velocity of υ during the storage process, let consider its velocity distribution as:

$$f(\upsilon) \sim e^{-m\upsilon^2/2k_bT}. \tag{4.35}$$

After some storage time t, the ith atom is no longer in its initial position $\boldsymbol{r}_i$, it moves to a new position $\boldsymbol{r}_i'$, which satisfies (here we approximate Eq. 3.35):

$$\boldsymbol{r}_i' = \boldsymbol{r}_i + \boldsymbol{\upsilon} t. \tag{4.36}$$

When the storage time increases, the displacements of different atoms will be different, and the accumulated phases will be different, too. Once the phase accumulation reaches $\varphi_i = 2\pi$, the phase of all atoms will be disordered, and the retrieval process will no longer be a state of interference enhancement. There will be no collective enhancement effect, as a result, the read-out photons will not emit along the specific direction with an expected probability. Deduced from Eq. 4.15, the retrieval efficiency is related to storage time:

$$\eta_r \propto \left| \frac{1}{N} \sum_{i=1}^{N} e^{i\Delta_k \cdot \upsilon t} \right|^2 = \left| \int d\upsilon f(\upsilon) e^{i\Delta_k \cdot \upsilon t} \right|^2 \propto e^{-t^2/\tau_d^2}, \tag{4.37}$$

where τ_d is the storage lifetime and related to atomic temperature T and spinwave's wave vector Δk.

$$\tau_d = \frac{1}{\Delta k \sqrt{k_b T/m}}. \tag{4.38}$$

When we perform the memory experiment, the angle between the direction of write beam and the direction of write-out photons is set to 3°, the atomic temperature is 10 μK. It can be calculated that the storage lifetime is about 78 μs, which is mainly limited due to the motion of spinwave.

4.2.5.1 Fidelity

In quantum memory, fidelity refers to the similarity between the retrieved quantum state and the stored quantum state. In the DLCZ quantum memory, there is actually no the process of storing incoming quantum states, it is by emitting write-out photons and then selecting the corresponding quantum state of the atoms, the atomic state at that time is called the quantum state stored in DLCZ quantum memory. This quantum state corresponds to the write-out photon. For a universal quantum memory, the density matrix corresponding to arbitrary quantum state that is stored in the quantum memory is ρ_i. After retrieving the quantum state from the memory, the density matrix is ρ_o, then the fidelity between two quantum states is defined as:

$$F = Tr\left(\sqrt{\sqrt{\rho_o}\rho_i\sqrt{\rho_o}}\right)^2. \tag{4.39}$$

For pure states, the definition of fidelity is also using the following formula:

$$F_{ps} = Tr(\rho_o\rho_i). \tag{4.40}$$

When the two quantum states are completely the same, the fidelity is 1. The higher the fidelity is, the better the performance of quantum memory is. For a completely unknown quantum state, it needs to perform quantum tomography [23] to acquire the full density matrix, however, the tomography process seems complex especially for multiple qubits in a large Hilbert space. Thus, in real experiments, if we just want to evaluate the overlap between a quantum state and a known target state (pure state like GHZ state), Eq. 4.40 is used more often, where only a few local measurements are needed [24].

4.2.5.2 Single-Photon Quality Function

During the experiment, we select the atomic state by detecting a write-out photon. In principle, the quantum memory should be retrieved with only one read-out photon, which has very good single-photon properties, but in fact, the write process is probabilistic. Although only one photon is detected, there are actually multiple excitations inside the atoms. Under this situation, whether the read-out photons is single photons will be affected. There may be multi-photon events in the read-out photons, and these events may be correlated to the undetected write-out photons, and a so-called pseudo-correlation between the detected write-out photons and multiple read-out photons exists, which results in that the quantum state of the read-out photons is different from the expected state. In order to assess the single-photon quality of the read-out photons and its multi-photon events, the single-photon quality function α (also called the second-order auto-correlation function) [25] can be defined. As shown in Fig. 4.3, it is a schematic diagram for measuring this function. Under

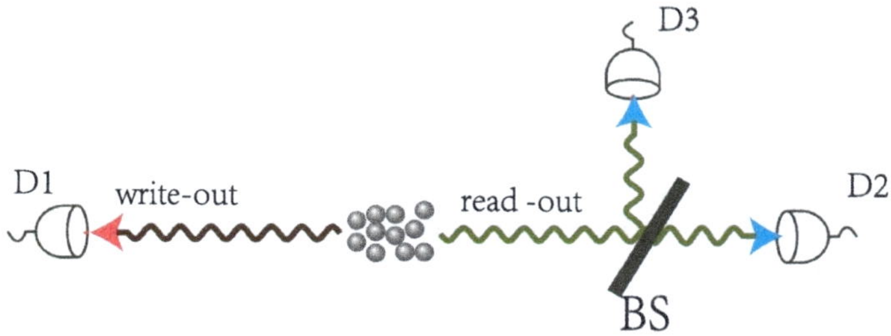

Fig. 4.3 Experimental schematic for measuring the single-photon quality function α

the condition that the write-out photon is detected, the read-out photon is divided into two paths with a beam splitter (BS), photons in each output port are detected by single photon detectors. Through the individual counts and coincidence counts of the detected events, the function α of read-out photon is calculated as:

$$\alpha = \frac{p_{23|1}}{p_{2|1}\, p_{3|1}}, \tag{4.41}$$

where $p_{23|1}$ is the coincident detected probability of read-out photons in detector D2 and D3 conditioned on the detection of write-out photons in detector D1, $p_{2|1}$ ($p_{3|1}$) is the detected probability of read-out photons in detector D1 (D2) conditioned on the detection of write-out photons in detector D1.

Second-Order Correlation Function

In our experiment, we can infer the polarization of the corresponding read-out photons by detecting the polarization of the write-out photons. Because polarizers, polarization beam splitters and other devices are often used in the experiment, these devices make the polarization of detected photons often fixed. It is not very convenient to still use fidelity to measure the correlation between write-out and read-out photons. Therefore, here we define a second-order correlation function to describe the relationship between write-out and read-out photons no matter what the polarizations of write-out and read-out photons are.

$$g^2 = \frac{\langle a_{wo}^\dagger a_{ro}^\dagger a_{ro} a_{wo} \rangle}{\langle a_{wo}^\dagger a_{wo} \rangle \langle a_{ro}^\dagger a_{ro} \rangle} = \frac{p_{wo,ro}}{p_{wo}\, p_{ro}}. \tag{4.42}$$

Here p_{wo}, p_{ro} is the measured probability of write-out photons and read-out photons in each experimental trial, $p_{wo,ro}$ is the measured probability of simultaneously detecting write-out photons and read-out photons in each experimental trial. If the two detected photons are completely un-correlated, then the probability of them being detected at the same experimental trial is the accidental coincidence probability $p_{wo}p_{ro}$, in this case, $g^2 = 1$. For the classic thermo-optical field, $g^2 = 2$. For $g^2 \gg 2$, it means that there is an obvious non-classical correlation between the two photons [26], which is enough to prove the quantum correlation between the read-out photons and the write-out photons, experimentally we often use this parameter to characterize that the memory is indeed a quantum memory rather than an ordinary memory for classical information. In our experiment, when the measured g^2 is not very large due to the signal-to-noise ratio and other reasons, Cauchy-Schwartz inequality can be also used to prove the non-classical correlation. For the classic light field, the inequality satisfies:

$$\frac{(g^2_{wo,ro})^2}{g^2_{wo,wo}g^2_{ro,ro}} \leqslant 1, \tag{4.43}$$

where $g^2_{wo,wo}$, $g^2_{ro,ro}$ are the auto-correlation function of write-out photons and read-out photons, respectively, which can be measured via sending the photons into a BS, and detecting the probability and coincidence probability of photons in the two outports after the BS. If the inequality is violated, the non-classical correlation between read-out and write-out photons can still be proved.

4.3 DLCZ Quantum Memory in the Free Space

4.3.1 Experimental Settings

As shown in Fig. 4.4a, all atoms are initially pumped into ground state $|g\rangle = |5^2S_{1/2}, F = 1\rangle$, the write beam coupling state $|g\rangle$ and $|5^2P_{1/2}, F' = 2\rangle$ is -10 MHz red detuned from the excited state. The position of cold atoms, i.e., the center of the MOT, is also where the beam waist of the write beam locates. The write beam with a beam waist of 300 μm interacts with the atoms and a write-out photon is scattered out through spontaneous Raman process, then a fiber coupler is used to collect the write-out photons in the direction with an angle of 3° apart from the write beam. The collected solid angle is calculated as $d\Omega = \pi(\frac{\lambda}{\pi w_{wo}})^2$, where the beam waist w_{wo} is chosen to 90 μm.

Since the write-out photon will be mixed with the photons via Rayleigh scattering, it will lead to the atoms returning back to its initial state and there will be no correlated atomic state even there is a write-out photon. To inhibit the influence of the Rayleigh process, we use a filter cavity (etalon) to filter out the Rayleigh scattered photons, whose frequency is distinguishable from the required write-out photons. The setup is shown in Fig. 4.4b. The etalon we use is a F-P cavity containing a flat surface and a concave surface, the radius of curvature of the etalon in the concave side is 100 mm, the length of the etalon is 6.35 mm, and the reflectivity of both sides of the etalon is 90%.

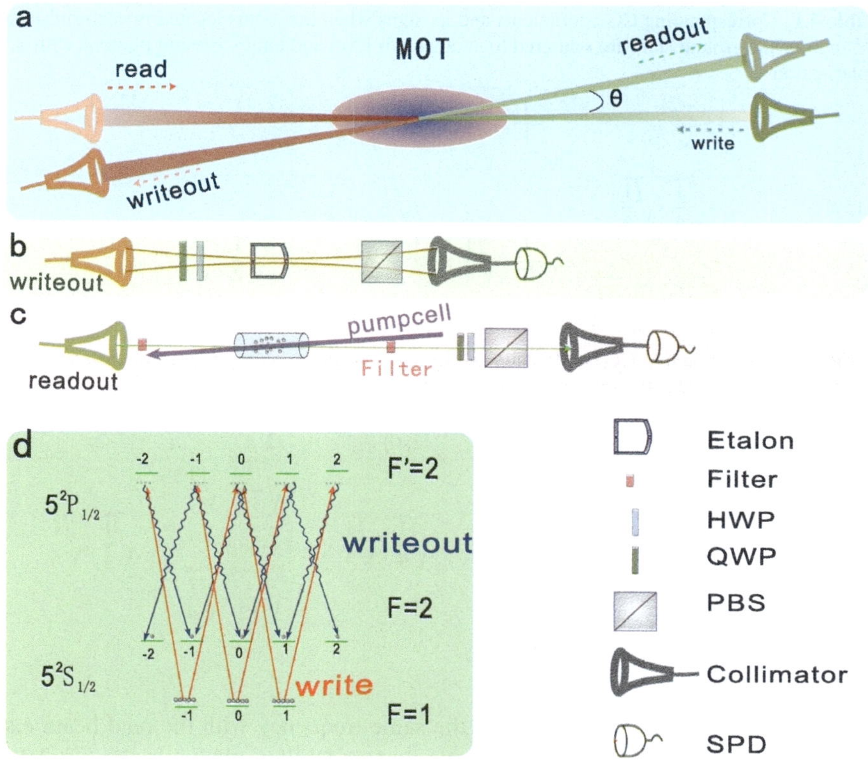

Fig. 4.4 Experimental setup and atomic energy levels for DLCZ quantum memory in free space. **a**, layout for write, read beam and write-out, read-out photons. The angle between write beam and write-out photons is $\theta \simeq 3°$. **b** filtering for write-out photons, etalon is used. **c**, filtering for read-out photons, atomic filtering cell is used. **d**, atomic energy levels in the quantum memory

Similarly, to filter the noise photons mixed in the read-out photons such as the leakage of read beam or Rayleigh scatted photons, we also utilize a filter.[1] Although an extra etalon can be placed in the read-out channel, the disadvantage is that the reflected write-out photons from the etalon in the read-out channel during the write process will be collected into the real write-out channel and thus result in extra noise photons in the write-out channel, here we use an atomic filtering cell instead of the etalon. As shown in Fig. 4.4c, the Rb cell is heated and pumped to $F = 2$ via a laser with transition of $F = 1 \rightarrow F' = 2$ at 780 nm (pumpcell laser mentioned in

[1] The filtering of read-out photons is mainly applied to improve the measured quantum correlation with write-out photons. It can be temporarily omitted when optimizing the state pumping and efficiency in the preliminary test. The main effect of non-filtering is that there will be more noise in the read-out photons, which may increase the accidental coincidences with the write-out photons, thus result in reducing the real quantum correlation between them, and affecting the fidelity of the entangled state in the followed experiment.

Table 4.1 Corresponding CG coefficients and its signs when the atoms located in each sub-level $|F, m_F\rangle$ of the ground state are scattered to another sub-level and emit write-out photons with σ^+ polarization

	$\|2,-2\rangle$	$\|2,-1\rangle$	$\|2,0\rangle$	$\|2,1\rangle$	$\|2,2\rangle$
$\|1,-1\rangle$	–	$-\sqrt{\frac{1}{12}}\cdot\sqrt{\frac{1}{4}}$	–	–	–
$\|1,0\rangle$	$-\sqrt{\frac{1}{4}}\cdot\sqrt{\frac{1}{6}}$	–	$-\sqrt{\frac{1}{4}}\cdot\sqrt{\frac{1}{4}}$	–	–
$\|1,1\rangle$	–	$-\sqrt{\frac{1}{12}}\cdot\sqrt{\frac{1}{4}}$	–	$-\sqrt{\frac{1}{2}}\cdot\sqrt{\frac{1}{6}}$	–

Table 4.2 Corresponding CG coefficients and its signs when the atoms located in each sub-level $|F, m_F\rangle$ of the ground state are scattered to another sub-level and emit write-out photons with σ^- polarization

	$\|2,-2\rangle$	$\|2,-1\rangle$	$\|2,0\rangle$	$\|2,1\rangle$	$\|2,2\rangle$
$\|1,-1\rangle$	–	$\sqrt{\frac{1}{2}}\cdot\sqrt{\frac{1}{6}}$	–	$\sqrt{\frac{1}{12}}\cdot\sqrt{\frac{1}{4}}$	–
$\|1,0\rangle$	–	–	$\sqrt{\frac{1}{4}}\cdot\sqrt{\frac{1}{4}}$	–	$\sqrt{\frac{1}{4}}\cdot\sqrt{\frac{1}{6}}$
$\|1,1\rangle$	–	–	–	$\sqrt{\frac{1}{12}}\cdot\sqrt{\frac{1}{4}}$	–

Chap. 3), thus the noise photons with the same frequency with the read beam can be absorbed and eliminated. Two dichroic mirrors ('Filter' shown in the Fig. 4.4c) are placed before and after the cell to let only the signal photons at 795 nm transmit and block the photons at 780 nm. The atomic cell allows the transmission efficiency around 90% for 795 nm photons and attenuation of 30 dB for 780 nm photons.

The write beam is set to H polarization, then the write-out photons would be V polarized, a polarization beam splitter (PBS) can be used for polarization filtering to get better signal-to-noise ratio. As shown in Fig. 4.4d, assuming that atoms are uniformly distributed in the sub-levels of $F = 1$, depending on the CG (ClebschGordan) coefficients of various transitions presented in Tables 4.1, 4.2, if the write beam is H polarized ($H = \frac{1}{\sqrt{2}}(\sigma^+ + \sigma^-)$), and the write-out photon is σ^- polarized, as the CG coefficient product signs are all '+', no extra phase will be introduced due to the transition. If the generated write-out photon is σ^+ polarized, the CG coefficient product signs are all '–', the transition will introduce additional phase π, so that the final state of two alternative polarized write-out photons is $\sigma^- - \sigma^+$, which corresponds to the V polarized write-out photons, orthogonal to the polarization of the write beam. One advantage of using this configuration is that it can use a PBS to further filter the signal photons to improve the signal-to-noise ratio. Similarly, when the read beam adopts V polarization, and the read-out photon will mainly be H polarization (Appendix C lists the comparison of the retrieval efficiency measured under the same conditions but different polarization settings in the experiment.)

4.3.2 Compensation of Environmental Magnetic Field

The existence of environmental magnetic field will affect the definition of the quantum axis and the storage lifetime of the quantum memory. Therefore, the ambient magnetic field (mainly the geomagnetic field) at the position of the atomic ensemble needs to be compensated to make the magnetic field zero. Here we utilize the change of the retrieval efficiency to optimize the magnetic field. If the magnetic field is not zero, the atomic energy level will undergo Zeeman splitting, and the phase of the evolution of different atomic energy levels over time is different, leading to a change of retrieval efficiency. By setting different storage time, the read-out photons will interfere due to the different accumulation phases. The direct result of the interference is that the retrieval efficiency will oscillate over time. As shown in Fig. 4.5a, the retrieval efficiency measured without magnetic field compensation changes with time. It can be seen that in addition to the exponential decay of retrieval efficiency in the normal storage lifetime, there exists an oscillation. The reduction of efficiency in the oscillation curve is the aforementioned process of interference cancellation in the read process. Since the magnetic field is zero under ideal conditions, the retrieval efficiency will not oscillate, so the storage time can be set in the middle point of the period of the first decrease in the retrieval efficiency (for example, 5 μs), and then adjust and optimize the current of three pairs of compensation coils in turn to maximize the retrieval efficiency. As shown in Fig. 4.5b, it is an example for the change of the retrieval efficiency with the current when the magnetic field is adjusted in one specific direction. Repeating the compensation process about three rounds, the time required for the retrieval efficiency decreasing zero due to the presence of the magnetic field will increase until it matches the decreasing envelope only resulted by the storage lifetime. When there is no oscillation, such as Fig. 4.5c, then it can be considered that the magnetic field compensation is sufficient.

4.3.3 Initialization of Atomic States

To start the quantum memory experiment, after trapping the atoms in the MOT, the atoms needs to be pumped to specific states. Take the Λ type three-level atom as an example, as the cooler laser couples a cycle transition of $|5^2S_{1/2}, F = 2\rangle \rightarrow |5^2P_{3/2}, F' = 3\rangle$, at the end of MOT loading, the atoms are almost in the state of s: $|5^2S_{1/2}, F = 2\rangle$, however, the experiment we planed needs the atoms located in the ground state g: $|5^2S_{1/2}, F = 1\rangle$, so the atom needs to be further pumped to the ground state. For this reason, a strong laser beam coupling s and excited state e: $|5^2P_{3/2}, F' = 2\rangle$ named as depumper laser (Fig. 3.5) is used to pump the atom to the ground state, the Hamiltonian of this depumper process can be written as [14]:

Fig. 4.5 Results for compensation of environmental magnetic field. **a**, the retrieval efficiency versus storage time without magnetic field compensation. **b**, When the magnetic field is compensated, the retrieval efficiency versus current of compensation coils. **c**, the retrieval efficiency versus storage time after magnetic field compensation.During the storage process, the atom is pumped to the atomic energy level of $F = 1$, and the polarization of write beam, read beam, write-out photon, and read-out photon is H, V, V, H, respectively

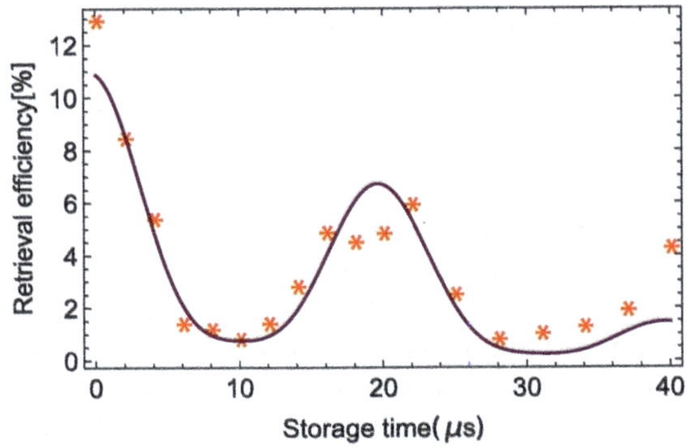

(a)

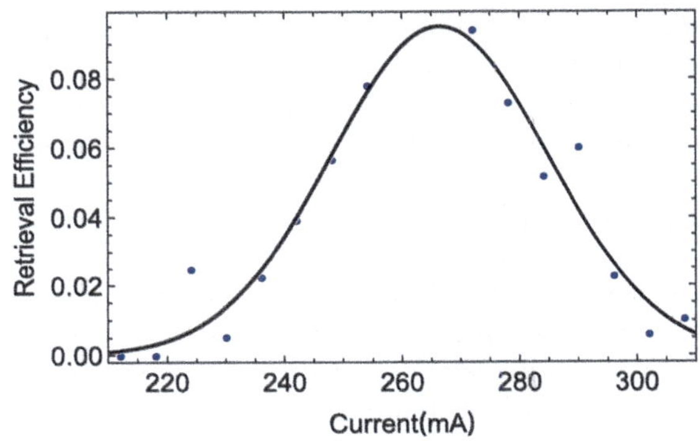

(b)

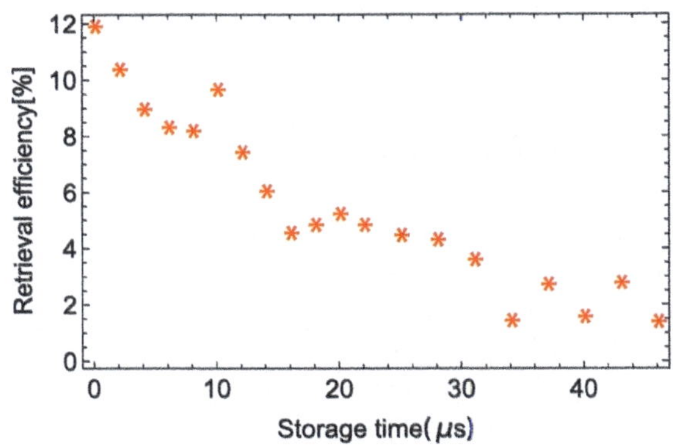

(c)

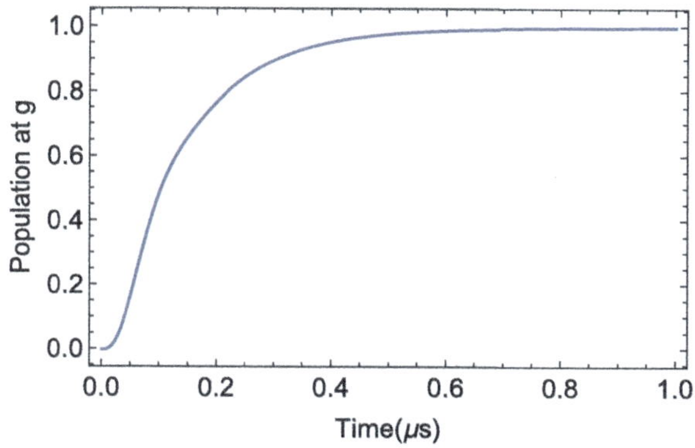

Fig. 4.6 State population versus pumping time in the depumper process

$$H = -\frac{\hbar}{2}\begin{pmatrix} 0 & 0 & 0 \\ 0 & 0 & \Omega_d \\ 0 & \Omega_d & 0 \end{pmatrix}, \tag{4.44}$$

where Ω_d is the Rabi frequency of the depumper light. Considering the decay of the excited states, the Lindblad equation [22] can be used to obtain the evolution of atomic density matrix ρ with time and the evolution should satisfy the equation:

$$\frac{d\rho}{dt} = \frac{1}{i\hbar}[H, \rho] + \frac{\Gamma_{eg}}{2}(2\sigma_{ge}\rho\sigma_{eg} - \sigma_{ee}\rho - \rho\sigma_{ee}) + \frac{\Gamma_{es}}{2}(2\sigma_{se}\rho\sigma_{es} - \sigma_{ee}\rho - \rho\sigma_{ee}). \tag{4.45}$$

Here Γ is the spontaneous decay rate, $\sigma_{ij} = |i\rangle\langle j|$ $(i, j = g, s, e)$.

In the experiment, the depumper laser is set to H polarized for interacting atoms in all sub-level states of $|5^2S_{1/2}, F = 1\rangle$, with a laser power of 2 mW and a beam waist of 1.8 mm. By solving Eq. 4.45, we can get the state population versus pumping time, as shown in Fig. 4.6.[2] From the figure, we can see that pumping time with 0.8 μs is enough to pump 99% of the atoms into the ground state g. However, as the Zeeman levels is degenerate without extra magnetic field, there will be a dark state between the degenerate sub-levels, which causes the atom not participating in the depumper process [27]. The frequency differences among excited states of D2 line is around the order of 100 or 200 MHz (Fig. 3.2), the depumper laser can couple other excited states with detuning. Even though the detuning is large, the influence of the strong laser can not be ignored. Therefore, the actual process of the entire depumping process is much more complicated than simply treating it as a three-level atom. In real situations, we set the pumping time more than 2 μs to obtain a satisfactory state initialization.

Since the subsequent experiments need to use the Zeeman sub-level of the atoms to generate the entanglement between the photonic polarizations and the internal state of the atoms, it is necessary to split the degenerate atomic energy level by

[2] An alternative method is to utilize the rate equation with Einstein coefficients. C. J. Foot. Atomic physics. Oxford University Press, 2005.

applying a bias magnetic field, and pumping the initial state of the atoms to a specified Zeeman splitted energy level. As shown in Fig. 4.7, the target state is $|g\rangle = |5^2S_{1/2}, F = 1, m_F = 0\rangle$, besides of the depumper laser, extra laser coupling $5^2S_{1/2}, F = 1 \rightarrow 5^2P_{1/2}, F' = 1$ is applied, which is the so called π-pump laser (resonant with the transition of $|6\rangle \rightarrow |8\rangle$). In the whole process, we can also use the Lindblad equation to evaluate the time it takes to initialize the atomic states. It is worth noting that once the atoms fall into the $|g\rangle$ state, since $|g\rangle \rightarrow |f\rangle$ is forbidden by electric dipole transitions, the atoms will stay in this state. As a result, we can pump all the atoms to this state. As the applied magnetic field is small, the splittings of Zeeman energy levels are small that they can be considered as degenerate. The pumping laser (Rabi frequency of depumper and π-pump is Ω_d and Ω_π, respectively) is considered as resonant with the excited states, then the Hamiltonian at the basis vectors of $\{6, g, 7, F = 2, 8, f, 9, F' = 2\} \times \{6, g, 7, F = 2, 8, f, 9, F' = 2\}$ can be written as:

$$H = -\frac{\hbar}{2}\begin{pmatrix} 0 & 0 & 0 & 0 & \Omega_\pi & 0 & 0 & 0 \\ 0 & 0 & 0 & 0 & 0 & 0 & 0 & 0 \\ 0 & 0 & 0 & 0 & 0 & 0 & \Omega_\pi & 0 \\ 0 & 0 & 0 & 0 & 0 & 0 & 0 & \Omega_d \\ \Omega_\pi & 0 & 0 & 0 & 0 & 0 & 0 & 0 \\ 0 & 0 & 0 & 0 & 0 & 0 & 0 & 0 \\ 0 & 0 & \Omega_\pi & 0 & 0 & 0 & 0 & 0 \\ 0 & 0 & 0 & \Omega_d & 0 & 0 & 0 & 0 \end{pmatrix}. \tag{4.46}$$

For the atoms pumped into excited states, they will decay to the ground states with decay rates Γ_{ij} ($i = \{8, f, 9, F' = 2\}$, $j = \{6, g, 7, F = 2\}$), which relate to the dissipative term in the equation.

$$\gamma_{dis} = \sum_{i,j} \frac{\Gamma_{ij}}{2}(2\sigma_{ji}\rho\sigma_{ij} - \sigma_{jj}\rho - \rho\sigma_{jj}) \tag{4.47}$$

The whole pumping process can be still described as:

$$\frac{d\rho}{dt} = \frac{1}{i\hbar}[H, \rho] + \gamma_{dis}. \tag{4.48}$$

Consider the initial boundary condition that atoms uniformly located in the five sub-levels of $F = 2$, i.e.,

$$\rho(0) = 0 \quad except \quad \rho_{44}(0) = 1. \tag{4.49}$$

As shown in Fig. 4.8, it simulates the change of atomic population during the entire pumping process versus the pumping time. It can be seen that the entire pumping process can basically pump all the atoms to $F = 1, m_F = 0$ within 2 μs, and then the atomic population will not change, however, the excited state in the experiment

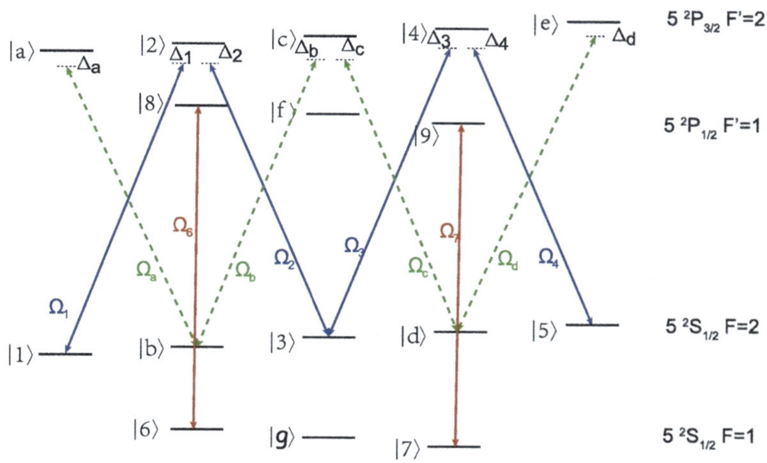

Fig. 4.7 The atomic energy levels and required pumping lasers for state initialization. The depumper laser with H polarization couples $|F = 2\rangle \rightarrow |F = 2'\rangle$. All atoms at sub-levels of $|F = 2\rangle, |F = 2'\rangle$ participate in the depumping process. The π-pump laser with π polarization couples $|F = 1\rangle \rightarrow |F = 1'\rangle$. Transition of $|1, 0\rangle \rightarrow |1', 0\rangle$ is forbidden by electric dipole transitions, atoms pumped into $|F = 1, m_F = 0\rangle$ will exit the pumping process and stay in the state

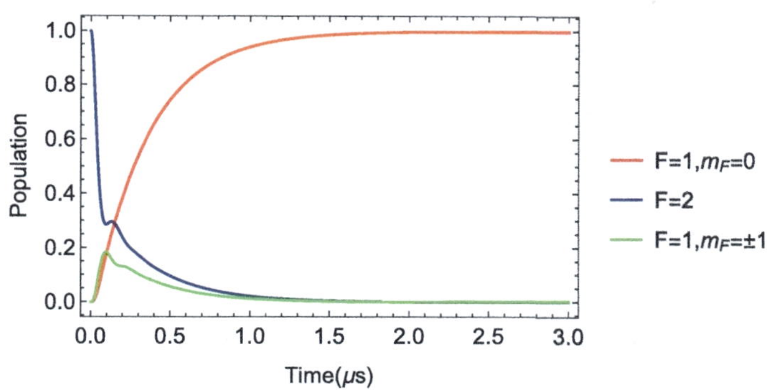

Fig. 4.8 Atomic state population versus the pumping time during the state initialization

has multiple energy levels, and the pumping laser may couple other energy levels under detuned conditions. The pumping procees in real experiment is even more complicated. In order to obtain a good pumping effect, the pumping laser is usually turned on for more than 3 μs to ensure that most of the atoms can be pumped into the specified energy level.

To assess the pumping effect, we examine whether the atoms still stay in $|F = 2\rangle$ or $|F = 1, m_F = \pm 1\rangle$. To verify whether the atoms are in state $|F = 2\rangle$, the method is simple. We apply lasers coupling $F = 2 \rightarrow F' = 2$, (for instance, the read laser), and then detect whether there exists scattered photons via the SPD (single photon detector). If the pumping process works, the detected events will stay at the dark count level of the SPD. On another hand, even if a small number of atoms are still at $|F=2\rangle$, in the DLCZ memory, these atoms have no collective enhancement effect in the retrieval process. The scattered photons are uniformly radiated along the entire space, and they are hardly collected into the read-out channel.

When examining whether atoms are still in $|F=1, m_F=\pm 1\rangle$, the π-pump can still be used following the aforementioned method. Here we verify it via the retrieval efficiency of quantum memories. When the bias magnetic field is applied, the atomic energy levels are no longer degenerate due to the existence of Zeeman splitting. If there are atoms on the sub-levels of ground state, then after applying the write beam (such as σ^+ polarization), if the write-out photon (σ^+) is detected, it means that there are atoms falling to one of the energy levels in $F=2, m_F=-1, 0, 1$. Atomic states at different energy levels will evolve and accumulate different phases due to splittings of energy levels (relative frequency is $\omega, 0, -\omega$). In the retrieval process, corresponding to the read-out photon at different initial states, the accumulated phase will be also different. The interference between these retrieval channels will cause the read-out photons from constructive interference to destructive interference, as a result, the retrieval efficiency will oscillate with the storage time, as shown in Fig. 4.9a. When the pumping process works well, most of the atoms will be in the $F=1, m_F=0$ state and the number of atoms in the $M_F=\pm 1$ state will be very small, whose contribution to the read-out photons can be ignored. The final retrieval efficiency will mainly come from the read-out photons of a single channel, and there will be no increase or decrease with the storage time, as shown in Fig. 4.9b. The small amplitude oscillation here comes from that impure light polarization and a small number of atoms are still in other energy levels. As mentioned in Sect. 4.2, the retrieval efficiency is related to the optical depth [28] and can be simply described as:

$$\eta_r = \frac{od}{od+1} = \frac{kN}{kN+1}, \tag{4.50}$$

where k is a constant, N is the atomic numbers, for the three initial atomic state $F=1, m_F=0, \pm 1$, the retrieval efficiency is related to the interference result from the contributions of the three states:

$$\eta_r = \left(\sqrt{\frac{kN_0}{kN_0+1}} + \sqrt{\frac{kN_1}{kN_1+1}}e^{i\omega t} + \sqrt{\frac{kN_{-1}}{kN_{-1}+1}}e^{-i\omega t}\right)^2. \tag{4.51}$$

The pumping efficiency is determined as:

$$\eta_p = \frac{N_0}{N_0+N_1+N_{-1}}. \tag{4.52}$$

By analyzing the results of Fig. 4.9b with π-pump laser and combining the intrinsic retrieval efficiency close to 0.4, it can be deduced that the atoms are almost at $m_F=0$.

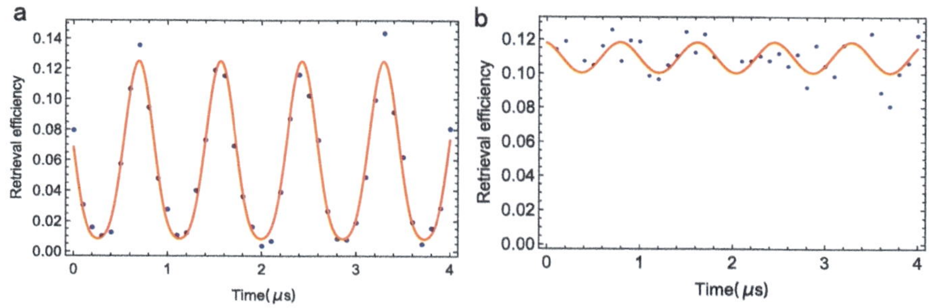

Fig. 4.9 Different retrieval efficiency in the different pumping process. **a**, without π-pump laser. **b**, with π-pump laser. It is fitted by $|Ae^{i\omega t}+B+Ce^{-i\omega t}|^2$. In **b**, the reason why some oscillations can still be seen is mainly due to the birefringence effect of the vacuum glasscells on the lasers. The polarization of the laser at the atomic position is not perfect. In the later entanglement experiments, we will optimize the polarization according to the entanglement visibility and the oscillation phenomenon will almost disappear under good pump conditions

The experimental pumping process needs to be optimized. There are two main optimization factors:

1, laser intensity and pumping time. There must be strong laser and enough pumping time to make the pumping process sufficient.[3]
2, polarization. For the depump process, Zeeman sub-levels are not involved, and the polarization requirements are relatively low. Usually, a depumper laser with approximately H polarization is used. For the π-pump process, the requirements for laser polarizations and incident angle are relatively strict. It must be H-polarized laser incident perpendicular to the direction of the magnetic field, and the polarization direction must be parallel to the magnetic field. In the experiment, the polarization plate is often used for calibration with a adjusting angle of 0.5°.

After the optimization of state preparation is completed, we measure the retrieval efficiency and the second-order correlation function over different storage time as shown in Fig. 4.10. The lifetime of retrieval efficiency is about 51 μs, mainly limited by the motion of the spinwave, that is, the spinwave vector caused by the angle between the write beam and the write-out photons in the experiment is non-zero and moves with the storage time. The lifetime of the second-order correlation function g^2 is 70 μs, which exceeds the classical limit 2, and the time to maintain quantum correlation is as long as 100 μs, which proves that the correlation between write-out and read-out photons is non-classical.

[3] In order to make full use of the existing lasers in the setup, avoid possible dark state phenomena, and shorten the pumping time, the read beam can also be turned on at the same time as a depumper laser in addition to the existing depumper laser.

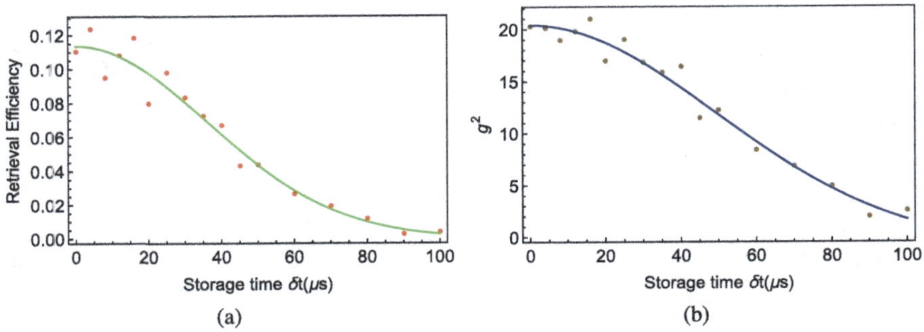

Fig. 4.10 Detected retrieval efficiency versus storage time after optimize the magnetic field and pumping process. **a**, retrieval efficiency. **b**, second-order correlation function g^2

4.3.4 *Accidental Coincidence Events*

In the experiment, the detections of write-out photons and read-out photons are not for deterministic single photons, just like two probabilistic single photon sources, even if they are not correlated, there will be a certain probability of coincidence detection. For our experiment, assume the excitation probability (that is, the detection probability of write-out photons) is p_w, the detection probability of the read-out photon is p_r, and the detection probability of the coincidences is p_{co}. The accidental coincidence probability between write-out and read-out photons is $p_w p_r$, then the net detected retrieval efficiency should be:

$$\eta_{net} = \frac{p_{co} - p_w p_r}{p_w} = \frac{p_{co}}{p_w} - p_r. \tag{4.53}$$

Correspondingly, the signal-to-noise ratio in the read process can be defined as:

$$SNR = \frac{p_{co}}{p_w p_r}. \tag{4.54}$$

It is the same with the g^2. It can be seen that g^2 also gives the signal-to-noise ratio of the quantum memory when measuring the non-classical correlation between write-out and read-out photons. Similarly, we can define the visibility as:

$$V = \frac{p_{co} - p_w p_r}{p_{co} + p_w p_r} = \frac{g^2 - 1}{g^2 + 1}. \tag{4.55}$$

When the read-out and write-out photons are one-to-one correlated, such as using a memory protocol based on Rydberg atoms, $V = 1$.

4.3.5 Multi-excitation Events

Due to the probabilistic write-out process of the DLCZ protocol in our experiment, when one atom is excited and emits a write-out photon, it is also possible that multiple atoms are excited and emit multiple write-out photons. Correspondingly, there will be multiple read-out photons later. This process is named multi-excitation process, it will make the measured retrieval efficiency differ from the real retrieval efficiency.

Assuming that the intrinsic excitation probability during the write-out process is η_w, the intrinsic retrieval efficiency is η_r, the overall transmission and detection efficiency of write-out photons is η_{dw}, and the overall transmission and detection efficiency of read-out photons is η_{dr}. Considering the process in which only one atom is excited, the probability of detecting a write-out photon in each experimental trial is:

$$p_w = \eta_w \eta_{dw}. \tag{4.56}$$

The probability to detect a read-out photon in the same experimental trial is:

$$p_r = \eta_w \eta_{dw} \cdot \eta_r \eta_{dr}. \tag{4.57}$$

When we consider the multi-excitations (here we consider two excitations), the probability to detect at least one write-out photon is:

$$p_{2w} = \eta_w^2 [1 - (1 - \eta_{dw})^2]. \tag{4.58}$$

And the probability to detect at least one read-out photon in the same experimental trial is:

$$p_{2r} = \eta_w^2 [1 - (1 - \eta_{dw})^2] \{\eta_r^2 [1 - (1 - \eta_{dr})^2] + 2\eta_r (1 - \eta_r) \eta_{dr}\}, \tag{4.59}$$

which includes two cases: two excitations are both successfully retrieved but only one photon is detected; only one of the two excitations is successfully retrieved and detected. The measured intrinsic retrieval efficiency thus can be calculated as:

$$\eta = \frac{p_r + p_{2r}}{(p_w + p_{2w})\eta_{dr}}. \tag{4.60}$$

In the cases of later experiments with cavity, $\eta_w = 0.02$, $\eta_{dw} = 0.5$, $\eta_r = 0.9$, $\eta_{dr} = 0.45$, the measured intrinsic retrieval efficiency is 0.916, increasing by less than 0.02 compared with 0.9.

4.4 DLCZ Quantum Memory with Ring Cavity Enhancement

From Sect. 4.2, it is known that the retrieval efficiency is largely affected by the collective enhancement effect, and the strength of the collective enhancement is related to the number of atoms. Due to the limitation of the experimental conditions, the number of atoms that can be trapped for the experiment is limited. In order to continue to improve the retrieval probability of emitting photons along the phase matching direction, we can use a cavity to couple the mode of read-out photons and cavity mode, which can help improve the probability of a read-out photon appearing in the cavity mode direction. As a result, a sufficiently high retrieval efficiency can be obtained. In the reference [29], Jonathan Simon et al. used a F-P cavity for the first time to enhance the retrieval efficiency. However, since the write-out and read-out photons are emitted both the forward and backward in the cavity, the generated spinwaves in two directions are mixed, and ultimately the actual detection efficiency and storage lifetime are not satisfying due to various filtering devices. In 2012, in the Ref. [30], Xiao-Hui Bao et al. improved the cavity design method and introduced a ring cavity to replace the F-P cavity for quantum memory experiments. Unlike the F-P cavity, which is a standing wave cavity, the ring cavity is a traveling wave cavity. In the experiment, the write-out and read-out photons can be separated along the two different directions of the ring cavity. The photons forward and backward will not be mixed together. The final intrinsic efficiency can reach 73%. Combined with the direction of gravity, collinear memory experiment was performed, which can improve the storage lifetime to 3.2 ms.

As mentioned in Chap. 2, inside a cavity with a finesse of $\mathcal{F}$, the probability of scattering photons by the atom along the cavity mode direction is enhanced by about $\frac{2\mathcal{F}}{\pi}$ (Eq. 2.43). The EIT process, where the generated read-out photon acts as the probe light and the read beam acts as the control light, is the same as the situation in free space (Sect. 4.2.3), which can be described by similar equations [31]:

$$
\begin{aligned}
\dot{E}_p &= -\kappa E_p + i g \rho_{eg} \\
\dot{\rho}_{eg} &= \left(-\frac{\Gamma_e}{2} + i\Delta_p\right)\rho_{eg} + \frac{i}{2} g E_p + \frac{i}{2}\Omega_c \rho_{sg} \\
\dot{\rho}_{sg} &= \left(-\frac{\Gamma_{sg}}{2} + i\delta\right)\rho_{sg} + \frac{i}{2}\Omega_c \rho_{eg} \\
E_{pout}(t) &= \sqrt{2\kappa} E_p(t)
\end{aligned} \quad . \tag{4.61}
$$

Here g represents the atom-photon coupling strength via cavity enhancement, κ is the decay rate of the cavity. When the read-out photons travels in the cavity round by round, it can be considered that the EIT process occurs multiple times through atoms in the cavity. The number of round trips of photons in the cavity is n (Eq. 2.30), each time it passes through the cavity mirror (transmittance $T = t^2$), a part of the photon will leave the cavity at the transmittance of T. In addition, the photon loss inside the cavity while passing through the atoms and instruments is $\gamma_a = Exp\{-kL\chi''\}$ and γ_s, respectively. $\gamma = \gamma_s + \gamma_a$. The output probability of a photon inside the cavity leaving the cavity can be approximately calculated as:

$$\eta_c \simeq \frac{T}{T+\gamma}. \tag{4.62}$$

4.4.1 Setup and Locking of the Ring Cavity

Before setting up the ring cavity, it is firstly necessary to calculate the cavity length to ensure that the ring cavity can be simultaneously resonant with both the write-out and read-out photons. In the previous experiments [30, 32], the write light, write-out photon, read light, and read-out photon are all resonant with the cavity, which means additional etalon and atomic filtering cell are still needed to filter noise photons caused by Rayleigh scattering like the cases in the free space. As a result, the overall photon detection efficiency is reduced. In order to make full use of the filtering function of the ring cavity, we set the write light to be red detuned to the excited state by 40 MHz, and the read light to be blue detuned to the excited state by 40 MHz, both off resonant with the cavity. As the write-out and read-out photons still need to be resonant with the cavity, it requires that the frequency difference between the write-out and read-out photons should be exactly an integer multiple of the free spectral range (here it has been assumed that waveplates have been used to compensate for the additional phase shift caused by the half-wave loss). Now the cavity length should obey the following equation:

$$\omega_{ro} - \omega_{wo} = n\frac{c}{L} = 6.834682\,\text{GHz} + 80\,\text{MHz}, \tag{4.63}$$

where we can calculate the cavity length as:

$$L = n \times 4.34 \quad \text{cm}. \tag{4.64}$$

Considering the real experimental layout, we finally choose the cavity length $L = 65.1$ cm wihch corresponds to a FSR of 461 MHz.

In addition to the frequency requirements for the cavity length, the mode-matching condition also requires the light to return to its original shape after one round trip in the cavity. The experimental write-out and read-out photons are all Gaussian modes with a beam waist of 90 μm, which will diverge with propagation, so it is necessary to place lenses in the cavity to recover the diverging modes to its original modes after a full round trip. As shown in Fig. 4.11a (Figure adapted from [33]), we use two convex lenses with focal lengths of $f_1 = f_2 = 250$ mm to realize the recovery of the Gaussian modes in the cavity. As shown in Fig. 4.11b, it is a schematic diagram of the transformation of Gaussian beam with a beam waist of w_0 through a lens with a

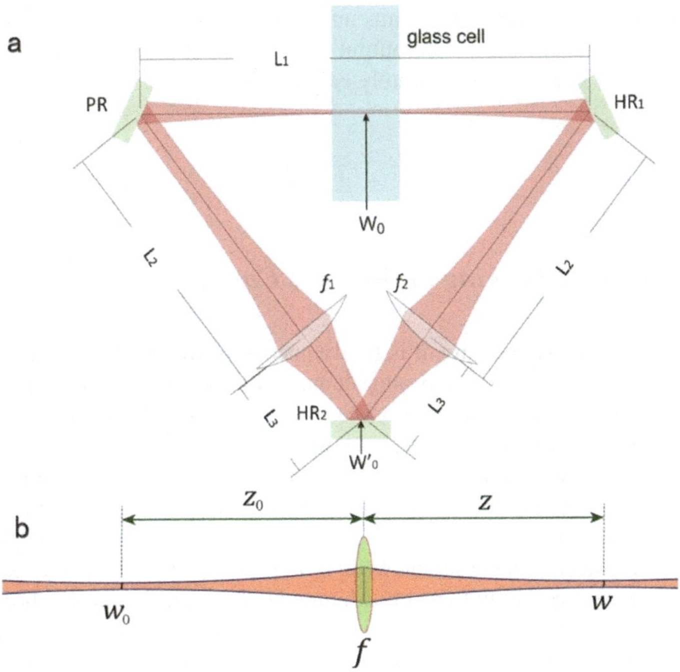

Fig. 4.11 Schematic diagram of the parameter design of the ring cavity and the transformation of Gaussian light passing through the lenses. Figure is reproduced from Ref. [33]

focal length of f. The transformed beam waist and the distance to the lens satisfy the following transformation equations:

$$\begin{aligned} z - f &= \frac{(z_0 - f)f^2}{(z_0 - f)^2 + z_R^2} \\ w &= w_0 \frac{f}{\sqrt{(z_0 - f)^2 + z_R^2}} \end{aligned}, \tag{4.65}$$

where $z_R = \frac{\pi w_0^2}{\lambda}$ is Rayleigh length. In the experiment, we use a symmetrical layout for the position of the lens in the cavity, so the position of the lens in Fig. 4.11a should meet the requirements:

The Gaussian beam with waist of w_0 passes through the lens with distance $L_1/2 + L_2$, and the Gaussian beam with waist w is obtained at the position of distance L_3.

In the end, we set the cavity modes and the beam waist to be approximately at the focal length of the lens, that is, $L_1/2 + L_2 = 250$ mm. Then the write-out photons passing through the lens can be viewed as a parallel light and then restore to its initial beam waist at the position of focal length after passing through the lens. Thus the beam waist can recover after every round in the cavity, a good cavity enhancement effect can be expected.

When the frequency and beam spot can have a good resonance enhancement effect with the cavity, the photons emitted by the atomic Zeeman sub-levels used in the experiment usually has circular polarization, and the $\sigma^{\pm}$ polarized light travels in the cavity will flip its polarization when it passes one round trip, and become $\sigma^{\mp}$ polarized light, which takes two round trips to recover. It will result in halving the FSR and affecting the effect of interference enhancement. In order to solve this problem, we use two quarter wave plates (QWP) and a half wave plate (HWP) to compensate for the phase difference between the two orthogonal lights. The placed structure is called a "sandwich" structure, where the waveplates are placed in turn. One QWP is fixed at 45°, then a freely adjustable HWP (assuming the fast axis angle is θ) is placed in the middle and another QWP at 45° is placed. The transformation after HWP and QWPs can be described by the following matrix:

$$U_{HWP}(\theta) = -i \begin{pmatrix} \cos(2\theta) & \sin(2\theta) \\ \sin(2\theta) & -\cos(2\theta) \end{pmatrix}, \tag{4.66}$$

$$U_{QWP}(\alpha) = \frac{1}{\sqrt{2}} \begin{pmatrix} 1 - i\cos(2\alpha) & -i\sin(2\alpha) \\ -i\sin(2\alpha) & 1 + i\cos(2\alpha) \end{pmatrix}. \tag{4.67}$$

Take $H = (1 \quad 0)^T$ and $V = (0 \quad 1)^T$ as examples, after passing through the three waveplates:

$$\begin{aligned} U_{QWP}\left(\frac{\pi}{4}\right) U_{HWP}(\theta) U_{QWP}\left(\frac{\pi}{4}\right) H &= e^{-i(\pi/2+2\theta)} H \\ U_{QWP}\left(\frac{\pi}{4}\right) U_{HWP}(\theta) U_{QWP}\left(\frac{\pi}{4}\right) V &= e^{i(\pi/2+2\theta)} V \end{aligned}. \tag{4.68}$$

It can be known that the polarizations of H and V recover after the three waveplates but a phase difference ($\phi = \pi + 4\theta$) between them can be acquired. It is clear that the relative phase between the two polarizations can be continuously adjusted by adjusting the HWP, which is very useful when adjusting the phase of the experimental light and the different phase between the phase-locked light in the phase-locked optical path. In view of the half-wave loss caused by the three-sided mirrors in the cavity, the relative phase π is introduced between them and the circular polarization is reversed. As a solution, by setting $\theta = \pi/4$, this extra phase can compensated so that the polarization will not change after the circularly polarized light goes around the cavity.

In order to set up the final experimental system (Fig. 4.12), the specific procedure of building the cavity can be briefly described as follows:

1, send the test write-out beam with the same frequency as the write-out photons from the opposite direction of write-out channel, initially set the polarization to H, place the three-sided cavity mirror to the preliminarily designed position, place a laser viewing card with a hole in the center in two positions and adjust

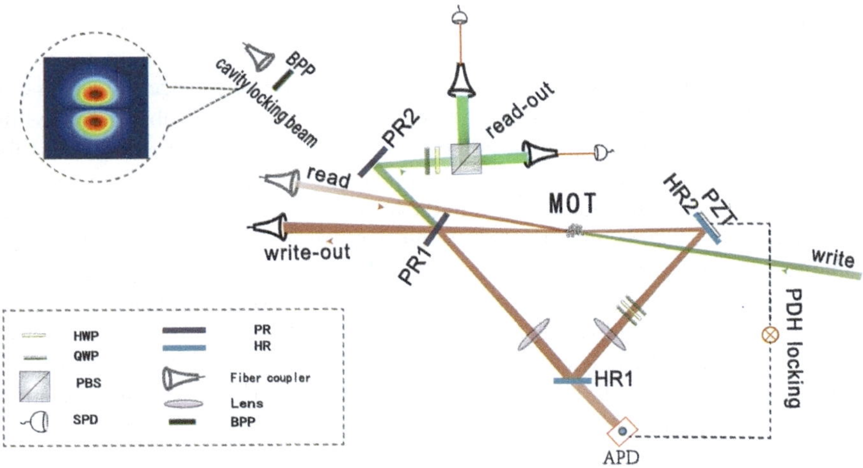

Fig. 4.12 Experimental layout for cavity enhanced quantum memories

the mirrors to ensure light can still pass through the centering hole after going around the cavity for the round trip.

2, scan the PZT on HR2, the light leaking from HR2 is received by a photo detector and observed by a oscilloscope. It should be able to display several transmission peaks during the scanning process.
3, preliminarily adjust the mirrors HR1 and HR2 to make the transmission peak during scanning the highest.
4, place the two lenses, during the placement process, try to wait until the transmission peak can be seen on the oscilloscope before fixing it.
5, send the test read-out beam with the same frequency as the read-out photons from the opposite direction of read-out channel, detect the light leaking from HR1 like the test write-out light.
6, adjust the one-dimensional translation stage under HR1 to change the cavity length, and ensure the two transmission peaks of test light on oscilloscopes can be completely overlapped. Finish the adjustment of cavity length.
7, jointly adjust the mirrors HR1 and HR2 to make the transmission peak as high as possible, and observe whether the line shape is smooth and consistent with the expected line shape (Lorenz curve).
8, adjust the up and down, left and right positions of the lenses, and also take the highest value of the transmission peak as the reference signal.
9, move the position of a lens back and forth in the cavity, repeat steps 7 and 8 for each movement, until the maximum value of the transmission peak no longer changes. Since the Rayleigh length of the waist of cavity modes in the experiment is about 3.2 cm, which is still much larger than the scale of the atomic ensemble, here we don't need to move both lenses to adjust, just moving one lens is sufficient.

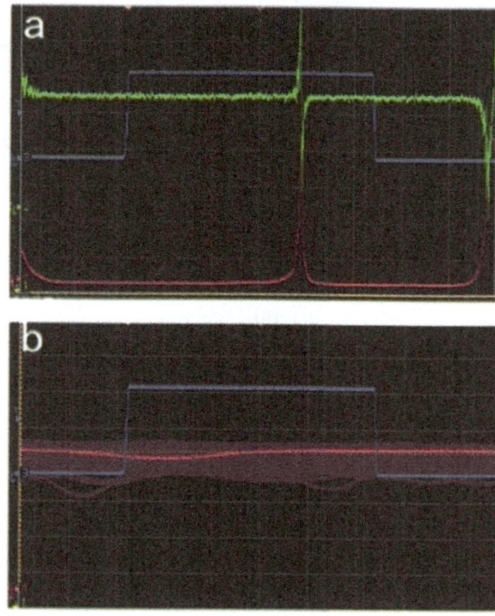

Fig. 4.13 Cavity locking signal. **a**, transmission signal and error signal while scanning the PZT. **b**, transmission signal for a while after locking the cavity

10, slightly move the lens on the fiber collimator of the write-out channel to see if the coupling with the cavity can be better.

11, place the "sandwich" structure compensation wave plates in the cavity, change the polarization of one of the test lights to circular polarization, adjust the angle of HWP so that the transmission peaks of different polarizations on the oscilloscope can overlap with each other. More subtle adjustments can be used by detecting the V polarized component in the light leaking from the cavity by setting the incident light H polarized. The angle of QWP with limited calibration accuracy can be slightly adjusted to ensure there is no V polarized component existing.

12, collect the test write-out light from the cavity locking beam channel, and then send the cavity locking beam back, detect the leakage light and send the detected signal to the FM-lock board [34], the analyzed error signal is sent to the PI board, and the output voltage is fed back and applied to the PZT, which is used to continuously and dynamically lock the cavity length to ensure that the photon phase in the cavity can be stabilized.

Figure 4.13a shows the transmission peak and error signal while scanning the PZT. To check the stability of the cavity after locking, we measure the drift of transmission signal for a while, as displayed in the rose red shade. It can be seen that the cavity transmission signal can be reduced by up to 20% during the locking process. After further optimizing the locking cavity parameters, the reduction is around 10%.

After completing the cavity locking, we test the quantum memory under the cavity enhancement according to the device shown in Fig. 4.12. By setting the write light and the read light red and blue detuned from excited state, respectively. The write-out and read-out photons are resonant with the cavity while the write and read light are

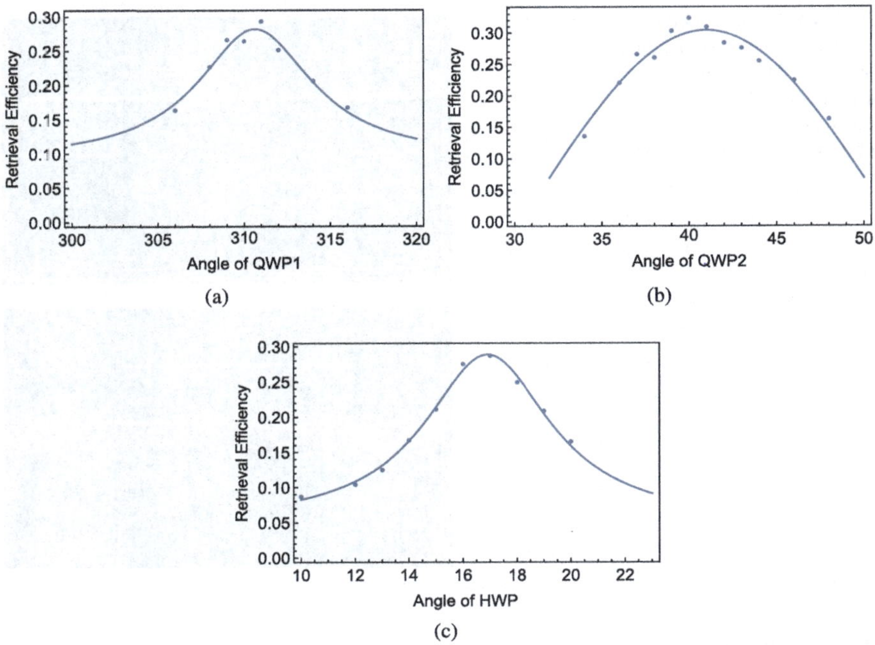

Fig. 4.14 Retrieval efficiency versus the angle of wave plates inside the cavity

off resonant with it. The noise photons generated by Rayleigh scattering process and light leakage are off resonant with the ring cavity. It can thus have an attenuation of 18 dB, so that we can completely eliminate the use of etalon and atomic filter cell in the experiment, which can further improve the transmission efficiency of the write-out and read-out photons.

4.4.2 Phase Compensation of Cavity Wave Plates

In order to further optimize the angle of the wave plates inside the cavity, we separately measured the retrieval efficiency versus the angle of each wave plate to verify the accuracy of the previous compensation. The results are shown in Fig. 4.14, which is almost the same as the previous compensation angle.

4.4.3 Frequency of Cavity Locking Beam and Read Beam

When adjusting the cavity, it is judged by the naked eye that both the test write-out and test read-out light resonate with the cavity. However, in real situations, the existence of atoms may cause phase shift, which causes the write-out and read-out photons not resonate with the cavity at the same time. To make the photons still resonate with the cavity simultaneously, we scan the frequency of cavity-locking beam firstly to make the excitation probability the highest, and then scan the frequency of read beam to make the retrieval efficiency the highest. As shown in Fig. 4.15a, if the frequency of the cavity locking beam is different, the excitation probability of generating write-out photons also changes under the influence of the cavity. Only when the write-out photons resonates with the cavity, the excitation probability is the highest, which determines the frequency of cavity locking beam. Then we fix the frequency of the write beam and the cavity locking beam, and scan the frequency of the read beam, only when the read-out photons is in resonance with the cavity, the retrieval efficiency is the highest, as shown in Fig. 4.15b, which determines the frequency of read beam. According to this setting, the write-out and read-out photons will maintain resonance with the cavity under real experimental conditions.

4.4.4 Noise from the Cavity Locking Beam

From the experimental layout shown in Fig. 4.12, it can be seen that the cavity locking beam is sent to the cavity via a PR in the read-out channel, thus the photons in the read-out channel can easily enter the write-out channel and be collected into the fiber for single photon detection. It will result in noise photons and affect the detection of the write-out photons. Even the locking beam is turned off during the write and read process, the early bright locking beam on the single-photon detector will also cause the increase of noise photons while detecting the write-out photons. As shown in Fig. 4.16, after loading the atoms in the MOT, we measure the probability of recording a noisy photon in each experiment trial when the intensity of the cavity locking beam varies.

The curve is nearly linear, indicating that only the noise photons caused by the cavity locking beam directly entering the detector during the whole process, and there is no photon generated by the interaction between the atoms and the cavity locking beam, which confirms that the cavity locking beam with one FSR difference from the atom-cavity resonance is far off resonance from the atomic states and can be viewed as a separate locking beam irrelevant to the atoms. Since the signal photon detected probability in the experiment is only on the order of 1%, in order to have a good signal-to-noise ratio, the requirement for cavity locking beam is high and the intensity as small as possible is expected. On the contrary, weak beam will inevitably affect the stability of the cavity locking. One possible way is to use an acousto-optic modulator as a switch, so that the cavity locking beam will not enter the

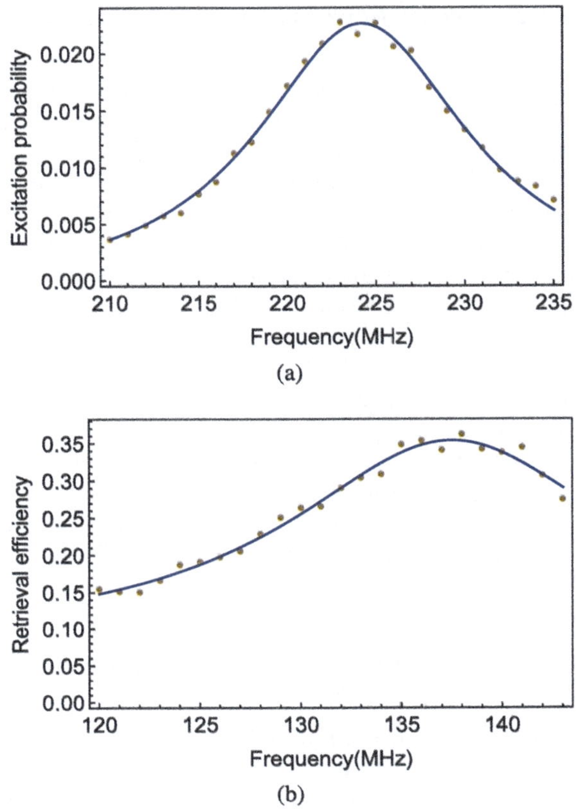

Fig. 4.15 Frequency of cavity locking beam and read beam. **a**, the change of excitation probability by setting the frequency of write beam and scanning the frequency of cavity locking beam. **b**, the change of retrieval efficiency while scanning the frequency of read beam

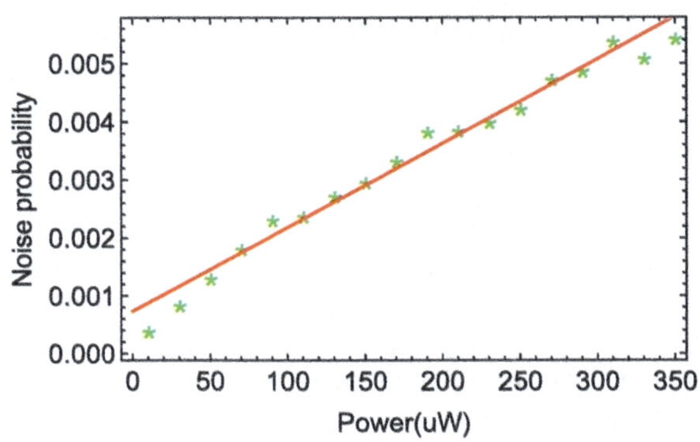

Fig. 4.16 The probability of detected noise photons from the cavity locking beam versus different intensity (measured time window of 100 ns)

detector when the cavity is locked, while during the write and read process, the signal photons passing through the acousto-optic modulator are diffracted and collected to the detector. This method solves the problem that early illumination on the SPD will have higher noise level. However, the use of AOM will sacrifice some signal photon transmission efficiency. To avoid this transmission loss, we adopt another method by changing the beam profile of cavity locking beam from Gaussian mode to HG_{10} mode (See the beam profile displayed in Fig. 4.12). The intensity in the center of HG_{10} mode is 0, hardly collected by the single-mode fiber that collects light with Gaussian modes. By doing this, it solves the problem of cavity locking beam entering the detector while retaining the transmission efficiency of signal photons.

4.4.5 Photonic Phase Shift Induced by Atoms

Each round the photon in the cavity passes through the atomic ensemble, there will be absorption and dispersion, which will cause it to acquire an additional phase and lead to a difference between the resonance frequency of the cavity with and without the atoms. For atoms, the susceptibility can be simply expressed as [35]:

$$\chi = \frac{i}{\Gamma + i \cdot 2\Delta}, \tag{4.69}$$

where $\Gamma = 5.73$ MHz is the natural line width, $\Delta = 40$ MHz is the detuning. Let $\delta = \frac{\Delta}{\Gamma}$, then $\chi_{Re} = \frac{2\delta}{1+4\delta^2}$ is relevant to the dispersion process and additional phase added to the photons, $\chi_{Im} = \frac{1}{1+4\delta^2}$ represents the absorption during the photon transmission in the atom ensemble. The photonic phase shift is:

$$\phi = od \cdot \chi_{Re}. \tag{4.70}$$

While scanning the frequency of cavity locking beam and read beam, the resonant frequency changes about 10 MHz compared with the frequency without loading cold atoms, we consider the FSR of 461 MHz, then can get that the phase changes about $\phi = \frac{10}{461} \cdot 2\pi = 0.14$, the optical depth can be calculated as:

$$od = \frac{\phi}{2\delta}(1 + 4\delta^2) = 2. \tag{4.71}$$

It is consistent with the measured od while absorption imaging and the can be used as another way to estimate the optical depth of the cold atoms.

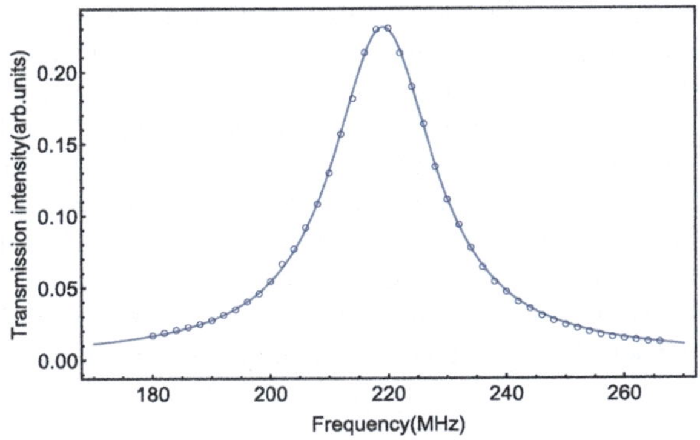

Fig. 4.17 The transmission signal obtained by scanning the test light frequency while the cavity is locked

4.4.6 *Cavity Finesse*

In our experiment, the cavity mirrors we used include one partial reflective mirror (PR1 with reflectivity $R = r^2 = 0.8$) and two high reflective mirrors (HR1 and HR2 with reflectivity larger than 0.99). A perfect cavity with these mirrors should have a finesse of:

$$\mathcal{F} = \frac{\pi\sqrt{r}}{1-r} = 28. \tag{4.72}$$

To measure the finesse of the cavity, there are two methods. One is to make use of the relationship between finesse, FSR and cavity linewidth ($\mathcal{F} = \nu_{FSR}/\delta\nu$). Through transmission peaks via scanning the PZT, the ratio between the FSR and linewidth can be obtained directly from the oscilloscope. This method is relatively simple and easy to implement, but the premise is that the displacement of the PZT changes linearly with the voltage during the scanning process. In fact, the relationship between displacement of PZT and the scanning voltage is not very linear, so this method is feasible for estimating the fineness, but not accurate enough for precise measurement. A better way is to lock the cavity, use the write-out channel to inject a beam of test light, monitor it behind the high-reflection mirror, and scan the frequency of the test light. With the scanning of the frequency, the resonance with the cavity will also vary. An example for the light leaking from the cavity while scanning its frequency is shown in Fig. 4.17.

By fitting the curve, the calculated finesse is about 23.5, smaller than 28. We attribute the decrease to the photon loss γ in the cavity, i.e., cavity loss, then the finesse is rewritten as:

$$\mathcal{F} = \frac{\pi\sqrt{r\sqrt{1-\gamma}}}{1-r\sqrt{1-\gamma}}. \tag{4.73}$$

It is calculated that the photon loss in the cavity is about 4.3%, mainly including the transmission loss through the glass cell ($\simeq$2.5%) and other optical elements such as lenses and waveplates. It is precisely due to the existence of cavity loss that high finesse cannot be achieved by selecting different partial mirrors. The higher the fineness, the more the average number of times the photons travels around the cavity. Although the interaction with the atoms is strengthened, the corresponding loss will also increase, which will eventually limit the photons that can leave the cavity and be collected.

4.4.7 Influence of Stability of Locked Cavity on Retrieval Efficiency

As shown in Fig. 4.13, even the cavity is locked, the ratio of intensity fluctuation of the resonant laser passing through the cavity can still reach 10%. It mainly caused by the instability of the locked cavity and will affect the cavity enhancement coefficient.

Considering the ring cavity used in our experiment, the cavity length is L, cavity loss is $\Gamma = 4.3\%$, the PR1 has a reflectivity of $R = r^2 = 80\%$ and a transmissivity of $T = t^2 = 20\%$. After the cavity is locked, we send a resonant laser into the cavity from the backward of the write-out channel and measure the intensity fluctuations of light leaking from the cavity, as the locked is not 100% stable, the intensity can decrease to 90% of its maximum value. The intensity of the transmitted laser is calculated as:

$$I = \frac{I_0}{1 + R \cdot (1 - \Gamma) - 2\sqrt{R \cdot (1 - \Gamma)} \cos \varphi}, \tag{4.74}$$

where $\varphi = \omega L/c$, I_0 represents the measured intensity while blocking the light going around the cavity. When the laser is resonant with the cavity, $\cos \varphi = 0$, $I_{max} = \frac{I_0}{(1-\sqrt{R \cdot (1-\Gamma)})^2}$. If the cavity is not stable, for example, the cavity length fluctuates and results in the decreasing of the intensity of transmitted laser to 90% of its maximum value. We can thus calculate the phase instability in the cavity as:

$$\cos \varphi = 0.998976. \tag{4.75}$$

Now we consider its influence on the retrieval efficiency. Assume the probability to retrieve a read-out photon on the phase-matched direction is p, after utilizing the cavity, the probability of read-out photons leaving the cavity in the specific direction is:

$$p_{out} = \frac{p(1 - \Gamma)t^2}{1 + (1 - \Gamma)R - 2\sqrt{(1 - \Gamma)R} \cos \varphi}. \tag{4.76}$$

In resonant cases,

$$\frac{p_{out}}{p}|_{\varphi=0} = 11.84. \tag{4.77}$$

As the change of cavity length caused by the instability of cavity locking will result in a phase different for photons traveling around the cavity, the ratio becomes:

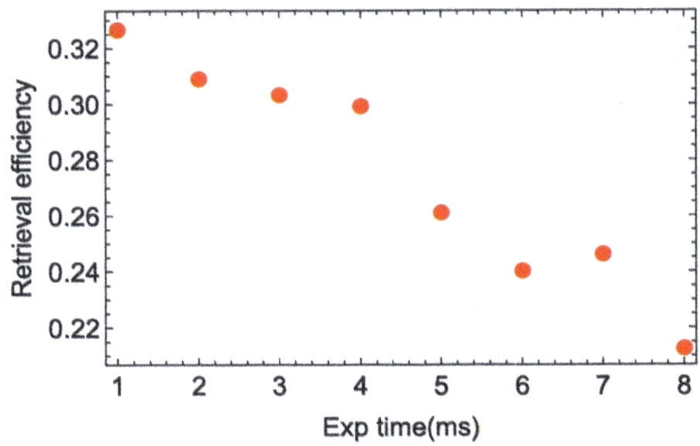

Fig. 4.18 The retrieval efficiency versus experimental times in each cold atomic loading

$$\frac{p_{out}}{p} = 10. \tag{4.78}$$

That means, the cavity enhancement coefficient has changed by about 10%, and the resulting change in retrieval efficiency is about 2.4%. Considering an overall measured retrieval efficiency, for instance, 35%,[4] the decrease due to the unstable cavity of the locking will be about 0.8%.

4.4.8 Influence of the Number of Experimental Trials on Retrieval Efficiency

In quantum memory experiments, the count rate is a critical parameter. How many effective events (mainly the coincidence counts between write-out and read-out photons) can be detected in one second directly affects the time-consuming of the experiment. For example, when loading the cold atoms, the longer the loading time, the more atoms can be captured, but at the same time the repetition frequency of the experiment will be reduced; after one loading, how many times of write and read process can be performed need to be optimized. If too many times are performed, the residual atoms becomes fewer and fewer, which will lead to the decrease of the retrieval efficiency and the signal-to-noise ratio, if the performed times is small, it will lead to a low experimental repetition rate, which is not a good choice for long-term experiments. Here we measure the change of retrieval efficiency by changing different experimental trials after each loading of the atoms, as shown in Fig. 4.18. The time sequence used is 18 ms to load and cool the atoms, and then the memory experiment will be carried out. A total of 5 μs is required for one write and read process.

[4] In our experiment, the experimental system is not always maintained in the best working status, sometimes the retrieval efficiency decreases and some experimental tests are still performed.

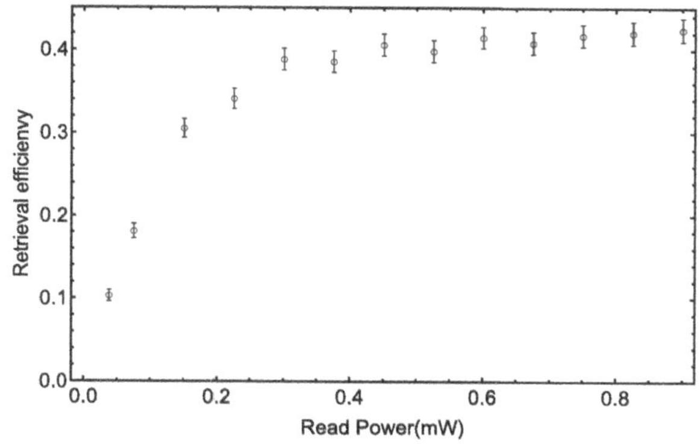

Fig. 4.19 Retrieval Efficiency versus intensity of read beam. The pulse width of read beam is 250 ns during the tests

It can be seen that the change of retrieval efficiency is not particularly obvious in the first 4 ms, and then it becomes lower and lower. The main reason is that the repeated pumping, write and read operations make the cold atoms less and less, and the atoms themselves leave the interaction areas with the lasers because of atomic motions.

4.4.9 Influence of Intensity of Read Beam on Retrieval Efficiency

During the read process, the intensity of the read beam directly affects the time required for the read-out photons to be retrieved and the overall retrieval efficiency. Low intensity will cause the read process to be too slow, and the pulse shape of the read-out photons will be too wide [30]. The main reason is that during the EIT process, there is a negative correlation between the read-out photon group velocity and the read beam intensity [14]. When the read beam is too weak, the photon group velocity is small, and it takes longer to pass through the atomic ensemble. This will cause the read-out photon pulse to widen. On the other hand, when the read beam intensity is too low, the excited collective atomic state may not be completely converted into a photon state, making the retrieval efficiency low. In the experiment, we tested the change of the retrieval efficiency with the intensity of read beam, as shown in Fig. 4.19.[5]

[5] The experimental data here is from the result of the later experiment where the initial state is at $|1, -1\rangle$, but the overall change trend is the same. In order to illustrate the effect of the influence factors on retrieval efficiency, they are placed in this subsection.

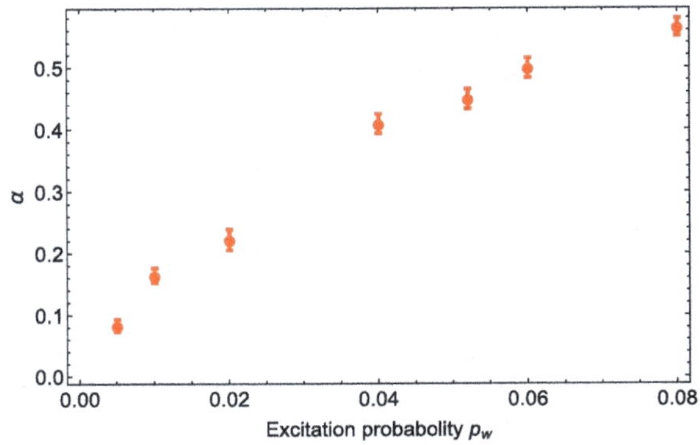

Fig. 4.20 Single photon quality of read-out photons versus excitation probability

4.4.10 *Influence of Excitation Probability on Single Photon Quality Function α*

In the DLCZ quantum memory described in this thesis, the generation process of write-out photons is probabilistic, which makes it possible to have 2, 3 or more write-out photons, and the probability of generating photons containing n write-out photons will be p_w^n where the excitation probability of emitting single write-out photon is p_w. Correspondingly, even if the detection of the write-out photon is used as the trigger, the read-out photon may contain multiple photons, and ideally we want that the read-out photons are perfect single photons. In order to describe the single photon quality of the read-out photons in the experiment, we measured the single photon quality function α under different excitation probabilities, as shown in Fig. 4.20. It can be seen from the figure that the lower the excitation probability, the better the single photon quality of the read-out photon, but at the same time the experimental count rate will be lower, which is not conducive to the experiment. After a tradeoff, we finally choose the excitation probability around 0.01.

4.4.11 *Influence of Excitation Probability on Second-Order Correlation Function* g^2

Since the retrieval efficiency is a fixed value, when the transmission and detection loss of write-out and read-out channels are similar, the measured probability of the read-out photons and measured probability of write-out photons (excitation probability) are nearly the same, i.e., $p_r \sim p_w$. The measured probability of coincidences between write-out and read-out photons should be the results of the measured probability of write-out photon multiplied by the retrieval efficiency plus accidental coincidence of write-out and read-out photons, which is:

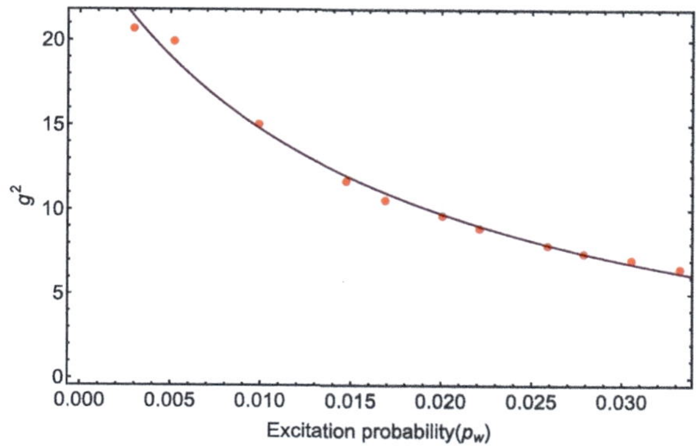

Fig. 4.21 Second-order correlation function g^2 versus excitation probability p_w

$$p_{co} = p_w\eta_{net} + p_r p_w \sim p_w\eta_{ro}. \tag{4.79}$$

Then the second-order correlation function satisfies:

$$g^2 = \frac{p_{co}}{p_w p_r} \propto \frac{p_w}{p_w \cdot p_w} \propto \frac{1}{p_w}. \tag{4.80}$$

It is inversely proportional to p_w, and the experimental measurement results are shown in Fig. 4.21.

4.5 DLCZ Quantum Memory with Atoms Initially Prepared in $|F = 1, m_F = -1\rangle$

Since our experiments of multiple cold atomic quantum memories utilize atoms initially prepared in the state of $|F = 1, m_F = -1\rangle$, here we will introduce some memory results in this case. The difference from the state preparation of the $|F = 1, m_F = 0\rangle$ is that the π-pump laser is no longer used here, instead, we use a σ^- pump laser with frequency the same with π-pump laser. The laser direction is also no longer the direction vertical to the bias magnetic field but parallel to the magnetic field. The polarization of the pump laser is σ^-. We use a BS to combine it with the write beam. Its polarization is opposite to the write beam and the direction is the same as the write beam. By doing so, we can prepare the atoms into $|F = 1, m_F = -1\rangle$ using the same time sequences like preparing the atoms into $|F = 1, m_F = 0\rangle$.

The storage process is the same as the DLCZ quantum memory described in the previous sections, the only difference is that the transition coefficients between used energy levels are different from that with initial state of $|1, 0\rangle$, as shown in Fig. 4.22. Previously, the CG coefficient amplitude ratio of the two write-out channels is $\sqrt{3} : \sqrt{2}$, and the CG coefficients of the two read-out channels are the same while right now the CG coefficient amplitudes of the two write-out channels will be the

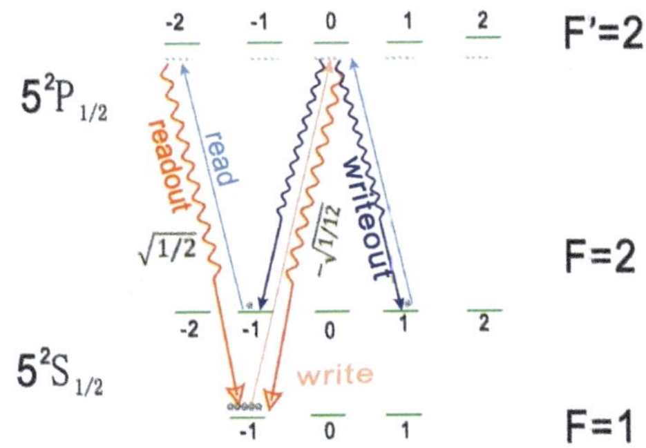

Fig. 4.22 Atomic energy levels with initial atomic state of $|1,-1\rangle$. Write: Light red solid arrow. Read: Light blue solid arrow. Write-out: Blue wavy line arrow. Read-out: Red wavy line arrow

same and the CG coefficient amplitudes of the read-out channel are quite different [36].

As mentioned in Sect. 4.2, the overall retrieval efficiency measured during the whole process, including the retrieval efficiency along the phase matching direction under collective enhancement (Eq. 4.19), cavity output efficiency (affected by the transmission efficiency of the EIT process when the read-out photon passes through the atoms (Eq. 4.32) and the loss γ during propagation in the cavity), transmission, collecting and detecting efficiency η_d outside the cavity. We write the transmission efficiency of photon passing through the atoms as (Eq. 4.32):

$$\eta_t = e^{-k_1 \cdot od}, \tag{4.81}$$

where k_1 is a constant. According to Chap. 2, the cavity enhancement coefficient is:

$$F = \frac{\eta_t(1-\gamma)T}{[1-\sqrt{\eta_t(1-\gamma)R}]^2}. \tag{4.82}$$

Then we can calculate the intrinsic retrieval efficiency with read-out photons emitting along the phase-matched direction as:

$$\eta_r = \frac{k_2 F \cdot od}{k_2 F \cdot od + 1}, \tag{4.83}$$

where k_2 is also a constant relevant to atomic energy levels. The cavity output efficiency is:

$$\eta_o = \frac{\eta_t(1-\gamma)T}{1-\eta_t(1-\gamma)R}. \tag{4.84}$$

Finally, the measured retrieval efficiency is:

$$\eta = \eta_r \eta_o \eta_d. \tag{4.85}$$

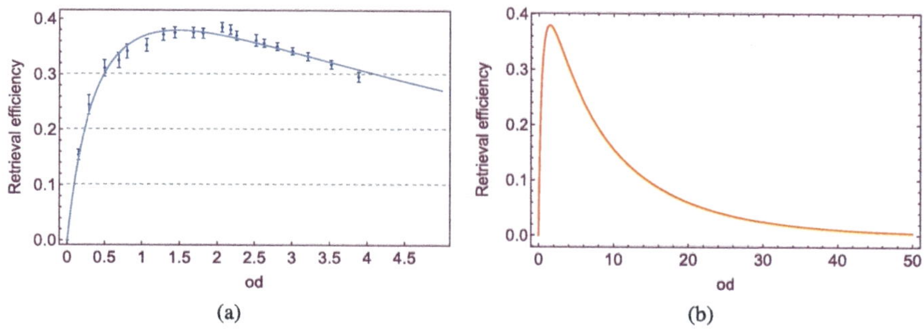

Fig. 4.23 Retrieval efficiency versus optical depth. **a**, measured efficiency versus *od*. In the experiment, *od* is usually set to around 1.5. During the test, the *od* is not measured one by one by the absorption imaging method, but is estimated by the approximate linear relationship between the excitation probability and *od*, so the data has some difference from the real *od*. The measured *od* is also related to the used transition line (here *od* is based on the measured *od* when the 795 nm wavelength laser with a CG coefficient of 1/4 is used), but the overall change trend is correct. **b**, the theoretical curve of the retrieval efficiency with increasing *od*, the signal transmission and collection efficiency (about 0.85) and the detection efficiency of 0.7 have been included

Fig. 4.24 Retrieval efficiency versus *od* with write-out photons σ^- polarization. The frequency of read beam is blue detuned from $F' = 2$ in the bottom curve and in the middle between $F' = 1$ and $F' = 2$ in the top curve

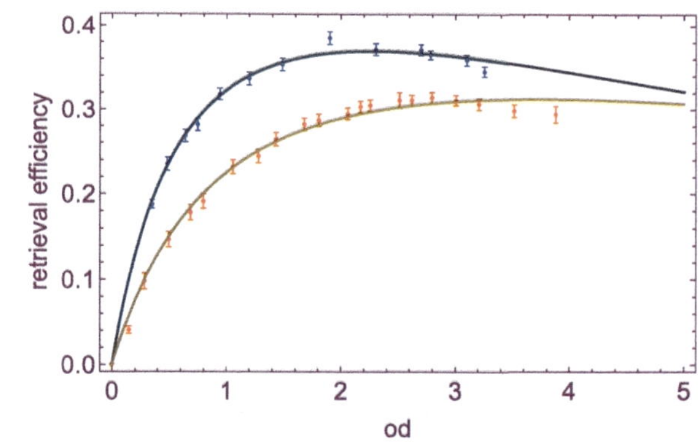

When we set the polarizations of write beam, write-out photons, read beam and read-out photons as σ^+ σ^+ σ^- σ^-, the measured result is shown in Fig. 4.23a. The solid line is the fitting curve through the above model, consistent with the experimental results.

It can be clearly seen from the figure that as the optical depth of the atoms increases, the retrieval efficiency will not continue to increase, but will begin to decrease. The main reason here is that after the optical depth becomes larger and larger, the loss will also increase, so that the output efficiency of read-out photons is reduced, and low retrieval efficiency is obtained.

When considering the write-out photon with σ^- polarization, due to the different transition coefficients between atomic energy levels, the frequency of the read beam when the read-out photon resonates with the cavity is different from the case where the write-out photon is σ^+ polarized. The experimental resonance frequency differs by about 15 MHz, and the maximum retrieval efficiency is only about 30%. In order to improve the retrieval efficiency under this read-out channel, we use the interference between the read-out channels to realize. The specific operation is to put the read beam in the center between $F' = 1$ and $F' = 2$, and the CG coefficients of read process through the two excited state energy levels is opposite in sign, with the detuning from the excited energy level "+" and "−", respectively, which is similar to Eq. 4.5 in the write process. It can be seen that the phase of the read-out photon through the two channels is proportional to $\frac{sign(CG)}{sign(\Delta)}$, so the phase of the read-out photon through the two energy levels is exactly the same. There will be a situation of constructive interference, so that the final retrieval efficiency can be improved compared with the situation of near resonant case. The measured result is shown in Fig. 4.24. From the experimental point of view, the channel with a smaller transition coefficient has lower retrieval efficiency. Here is just a method to improve the retrieval efficiency. In the following entanglement experiments, due to the limitation of experimental conditions, we did not use this method to read. After different write process, the same read beam is used, and the frequency of read beam is optimized according to the standard of high retrieval efficiency in the read-out channel with higher retrieval efficiency.

By further optimizing the intensity splitting ratio of the cooler laser, it is possible to decrease the loss of atoms over time in repeated experiments after each loading of cold atoms, thereby reducing the impact of the reduction of the atomic optical depth and the change of cavity resonant conditions on the retrieval efficiency.

4.6 Summary

In this chapter, I mainly introduce the principles and experiments of the DLCZ quantum memory in detail, and describe in detail the experimental factors that affect the retrieval efficiency and how to optimize it step by step to obtain high retrieval efficiency. In the end, the final detection value of the retrieval efficiency we can obtain is up to 0.42, which corresponds to that the intrinsic retrieval efficiency inside the atoms has exceeded 0.9. It is the best result in the current quantum memory based on the cold atom ensembles. Our work will provide very favorable conditions for building quantum networks with multiple cold atomic quantum memories.

References

1. Lvovsky AI, Sanders BC, Tittel W (2009) Optical quantum memory. Nat Photonics 3(12):706
2. Simon C, Afzelius M, Appel J, Boyer de La Giroday A, Dewhurst SJ, Gisin N, Hu CY, Jelezko F, Kröll S, Müller JH, et al (2010) Quantum memories. Eur Phys J D 58(1):1–22
3. Bussières F, Sangouard N, Afzelius M, de Riedmatten H, Simon C, Tittel W (2013) Prospective applications of optical quantum memories. J Mod Opt 60(18):1519–1537

4. Sangouard N, Simon C, De Riedmatten H, Gisin N (2011) Quantum repeaters based on atomic ensembles and linear optics. Rev Mod Phys 83(1):33
5. Kimble HJ (2008) The quantum internet. Nature 453(7198):1023
6. Jiang L, Taylor JM, Sørensen AS, Lukin MD (2007) Distributed quantum computation based on small quantum registers. Phys Rev A 76(6):062323
7. Appel J, Windpassinger PJ, Oblak D, Hoff UB, Kjaergaard N, Polzik ES (2009) Mesoscopic atomic entanglement for precision measurements beyond the standard quantum limit. Proc Natl Acad Sci 106(27):10960–10965
8. Barrett SD, Rohde PP, Stace TM (2010) Scalable quantum computing with atomic ensembles. New J Phys 12(9):093032
9. Hsiao YF, Tsai PJ, Chen HS, Lin SX, Hung CC, Lee CH, Chen YH, Chen YF, Ite AY, Chen YC (2018) Highly efficient coherent optical memory based on electromagnetically induced transparency. Phys Rev Lett 120(18):183602
10. Cho Y-W, Campbell GT, Everett JL, Bernu J, Higginbottom DB, Cao MT, Geng J, Robins NP, Lam PK, Buchler BC (2016) Highly efficient optical quantum memory with long coherence time in cold atoms. Optica 3(1):100–107
11. Guo J, Feng X, Yang P, Yu Z, Chen LQ, Yuan CH, Zhang W (2018) High-performance Raman quantum memory with optimal control. arXiv:1804.10763
12. Vernaz-Gris P, Huang K, Cao M, Sheremet AS, Laurat J (2018) Highly-efficient quantum memory for polarization qubits in a spatially-multiplexed cold atomic ensemble. Nat Commun 9(1):363
13. Duan LM, Lukin MD, Cirac JI, Zoller P (2001) Long-distance quantum communication with atomic ensembles and linear optics. Nature 414(6862):413
14. Fleischhauer M, Imamoglu A, Marangos JP (2005) Electromagnetically induced transparency: optics in coherent media. Rev Mod Phys 77(2):633
15. Phillips DF, Fleischhauer A, Mair A, Walsworth RL, Lukin MD (2001) Storage of light in atomic vapor. Phys Rev Lett 86(5):783
16. Scully MO, Zubairy MS (1999) Quantum optics
17. Duan L-M, Cirac JI, Zoller P (2002) Three-dimensional theory for interaction between atomic ensembles and free-space light. Phys Rev A 66(2):023818
18. Straßel T (2008) Quantum memory with atomic ensembles of rubidium and single photons for long distance quantum communication. PhD thesis, University of Heidelberg
19. Zhao B (2008) Robust and efficient quantum repeater with atomic ensembles and linear optics. PhD thesis, University of Heidelberg
20. Saffman M, Walker TG, Mølmer K (2010) Quantum information with Rydberg atoms. Rev Mod Phys 82(3):2313
21. Bao X-H (2010) Quantum information with entangled photons and cold atomic ensembles. PhD thesis, University of Heidelberg
22. Breuer HP, Petruccione F, et al (2002) The theory of open quantum systems. Oxford University Press on Demand
23. James DFV, Kwiat PG, Munro WJ, White AG (2001) Measurement of qubits. Phys Rev A 64(5):052312
24. Gühne O, Tóth G (2009) Entanglement detection. Phys Rep 474(1–6):1–75
25. Grangier P, Roger G, Aspect A (1986) Experimental evidence for a photon anticorrelation effect on a beam splitter: a new light on single-photon interferences. Eur Lett (EPL) 1(4):173–179 Feb
26. Kuzmich A, Bowen WP, Boozer AD, Boca A, Chou CW, Duan L-M, Kimble HJ (2003) Generation of nonclassical photon pairs for scalable quantum communication with atomic ensembles. Nature 423(6941):731
27. Michael, Spitchtinger Jan (2014) Dark-state formation during ground state transfer with optical pumping: [bachelor thesis]. University of Heidelberg
28. Gorshkov AV, André A, Lukin MD, Sørensen AS (2007) Photon storage in λ-type optically dense atomic media. ii. free-space model. Phys Rev A 76(3):033805

29. Simon J, Tanji H, Thompson JK, Vuletić V (2007) Interfacing collective atomic excitations and single photons. Phys Rev Lett 98(18):183601
30. Bao X-H, Reingruber A, Dietrich P, Rui J, Dück A, Strassel T, Li L, Liu N-L, Zhao B, Pan J-W (2012) Efficient and long-lived quantum memory with cold atoms inside a ring cavity. Nat Phys 8(7):517
31. Gorshkov AV, André A, Lukin MD, Sørensen AS (2007) Photon storage in λ-type optically dense atomic media. i. cavity model. Phys Rev A 76(3):033804
32. Yang S-J, Wang X-J, Li J, Rui J, Bao X-H, Pan J-W (2015) Highly retrievable spin-wave-photon entanglement source. Phys Rev Lett 114(21):210501
33. Dück A (2009) Ring cavity for quantum memory with atomic ensembles. Master's thesis, University of Heidelberg
34. Jiang X (2009) The laser frequency stabilizing and phase locking on quantum memory based on cold atoms. PhD thesis, University of Science and Technology of China
35. Campbell GK, Leanhardt AE, Mun J, Boyd M, Streed EW, Ketterle W, Pritchard DE (2005) Photon recoil momentum in dispersive media. Phys Rev Lett 94(17):170403
36. Steck DA (2001) Rubidium 87 D line data

Chapter 5
Entanglement of Three Cold Atomic Ensembles

After achieving the highly retrievable quantum memory, in this chapter, I will mainly introduce the influence of magnetic field, photon polarization, and multiple excitation cases on the fidelity of photon-atom entanglement and optimize these factors to achieve high fidelity photon-atom entanglement. Furthermore, we analyze and solve the influence of pulse shape, frequency and other factors on photon-photon interference, and realize high-quality atom-atom entanglement. Finally, we generate the entanglement of three cold atomic ensembles and realize the basic building block of a two-dimensional quantum repeater [1], paving the way for the realization of a quantum network.

5.1 Background

In order to connect different quantum network nodes, entanglement between flying photons and stationary atoms is the best resource. As atoms can be placed as quantum memory in each node to store and manipulate quantum information, while photons can be transmitted to other places through the network and interfere with each other, so that the entanglement of the quantum memories at the quantum network nodes can be generated, and finally the construction of quantum network is hopeful. Therefore, obtaining a high-quality source of entanglement between photons and atoms has become the first urgent problem that needs to be solved. On the one hand, the generation and verification rate of the entanglement source must be high enough, and on the other hand, the fidelity of the entanglement source must be high enough. The generation rate of the entanglement source can be increased by optimizing the time sequences and excitation probability. For the atomic state that needs to be verified, we usually convert it into a photonic state for detection and verification, so the retrieval efficiency of the stored atomic state is particularly important, which has

B. Jing, *Quantum Network with Multiple Cold Atomic Ensembles*, Springer Theses,
https://doi.org/10.1007/978-981-19-0328-1_5

been introduced in detail in the previous chapter. The main point of this chapter is to study the realization of high-fidelity photon-atom entanglement and entanglement of multiple cold atomic ensembles.

5.2 Entanglement Between Single Photons and Cold Atomic Ensembles

In the previous chapter, I introduced the experimental implementation of DLCZ quantum memory. The initial state of the atoms is prepared at $|g\rangle = |F = 1, m_F = 0\rangle$, after a 'write' operation, if events with two or more write-out photons are ignored and only the case of one write-out photon is considered, there will be one atom being excited to $|s\rangle = |F = 2, m_F = 0\rangle$. The state between the write-out photon and the atom can be then written as:

$$|\Psi\rangle_{ap} = |g\rangle_a|0\rangle_p + \sqrt{p}|s\rangle_a|1\rangle_p, \tag{5.1}$$

where p is the excitation probability. $|\Psi\rangle_{ap}$ is an entangled state, which describes the correlations between whether there is a write-out photon and whether the atom is Raman scattered. When the atom is Raman scattered, the Fock state of write-out photon is $|1\rangle_p$, when the atom is not scattered, the atom stays in its initial state and the Fock state of write-out photon is $|0\rangle_p$. In 2004, D. N. Matsukevich et al. reported the first demonstration of entanglement between photons and atoms [2]. They split a single write beam into two and send the two beams into two cold atomic ensembles, the polarizations of generated write-out photons from the two ensembles are different, by overlapping the photons with a polarization beam splitter, one can get entanglement between polarizations of write-out photon and which atomic ensembles are excited. In 2005, Chou Chin-Wen et al. used Fock state entanglement and interfere the write out photons from two ensembles with a BS, then realized the entanglement between two atomic ensembles ($|sg\rangle + |sg\rangle$) for the first time through a single photon detection method [3]. Their results represent the quantum repeater node described in Ref. [4], however, as this kind of entanglement makes use of whether there is a photon and only one photon is detected, in long distance quantum communication, more and more repeater nodes will result in increase of vacuum component of the state and affect the final communication length and fidelity [5]. To avoid the influence, two-photon detection method is developed. One of the typical schemes is experimentally demonstrated in 2007 by Shuai Chen et al. [6]. They make use of the different directions of write-out photons, which is correlated with the spatial modes of collective excitation. By designing the experiment, the polarizations of write-out photons is correlated with the spatial modes of atomic excitation, thus the vacuum state can be moved and entanglement between polarizations of photons and spatial modes of atomic excitations can be generated. In 2015, Shengjun Yang et al. further improve the experiment and directly use the different Zeeman sub-energy levels of atoms to

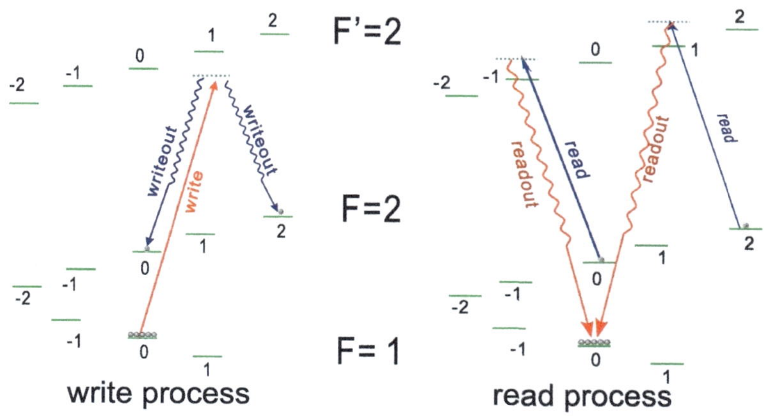

Fig. 5.1 Atomic energy levels of write and read beam for generating photon-atom entanglement

realize the entanglement between photon polarizations and different atomic energy levels [7]. The photon-atom entanglement used in this thesis is also generated in this way.

As shown in Fig. 5.1, the atomic states are initially prepared into $|1,0\rangle$, a write beam with σ^+ polarization is sent to the atoms, through Raman scattering, a write-out photon will be generated together with a collective excitation for atoms. When the write-out photon is σ^+ polarized, the excited atom will fall into state $|s_1\rangle = |2,0\rangle$, when the write-out photon is σ^- polarized, the excited atom will fall into state $|s_2\rangle = |2,2\rangle$, considering the CG coefficients during the transitions, i.e.,

$$\begin{aligned} CG_1(|2,0\rangle \rightarrow |2',1\rangle) &= \sqrt{\frac{1}{4}} \\ CG_2(|2,2\rangle \rightarrow |2',1\rangle) &= -\sqrt{\frac{1}{6}}, \end{aligned} \tag{5.2}$$

the generated entanglement between the photonic polarizations and atomic spin states can be written as:

$$|\Psi\rangle_{ap} = \sqrt{3/5}|s_1\sigma^+\rangle + \sqrt{2/5}e^{i\varphi(t)}|s_2\sigma^-\rangle. \tag{5.3}$$

Here $\varphi(t) = \pi + 2\mu Bt/\hbar$ is the phase difference accumulated by different Zeeman energy levels with storage time. To confirm the entanglement, CHSH inequality can be measured [8], here we simply measure the visibility under different measurement bases to verify whether the photon and the atom are entangled. In order to measure the atomic state, based on the efficient read process and the one-to-one correspondence between the read-out photon and the internal state of the atom, we can convert the spin state of the atom into the polarization state of the read-out photon by read beam.

By measuring the correlations between write-out and read-out photons, we can verify the photon-atom entanglement. By adjusting the storage time and setting the phase φ to 0, the entangled state between write-out and read-out photon can be written as:

$$|\Psi\rangle_{wr} = \sqrt{3/5}|\sigma^+\sigma^-\rangle + \sqrt{2/5}|\sigma^-\sigma^+\rangle, \tag{5.4}$$

where the first qubit corresponds to the write-out photon while the second qubit corresponds to the read-out photon. Let define $R/L := \sigma^+/\sigma^-$. For simplicity, we perform a unitary operation on the two photons with waveplates and convert the circular polarization into H/V polarization, the entangled state can be rewritten as:

$$|\Psi\rangle_{wr} = \sqrt{3/5}|HH\rangle + \sqrt{2/5}|VV\rangle. \tag{5.5}$$

By setting different polarization bases for measurement and measuring the write-out and read-out photon counting probability and the coincidence counting probability p_{wi}, p_{rj}, p_{ij} ($i, j = \{H, V\}$ represents the two cases of polarization for write-out and read-out photons), the visibility is defined as:

$$V_{HV} = \frac{p_{HH} + p_{VV} - p_{HV} - p_{VH}}{p_{HH} + p_{VV} + p_{HV} + p_{VH}}. \tag{5.6}$$

To confirm the two photons are indeed entangled, the correlations under superpositional bases also need to be measured, i.e., bases at $R/L = H \pm iV$ and $D/A = H \pm V$. The entanglement visibility can be also defined as:

$$\begin{aligned} V_{RL} &= \frac{p_{RL} + p_{LR} - p_{LL} - p_{RR}}{p_{LR} + p_{RL} + p_{LL} + p_{RR}} \\ V_{DA} &= \frac{p_{DD} + p_{AA} - p_{DA} - p_{AD}}{p_{DD} + p_{AA} + p_{DA} + p_{AD}} \end{aligned}. \tag{5.7}$$

It can be seen that in ideal cases, there should be no coincidences of HV and VH, the visibility is 1. Here we first explain what non-correlated items will be affected by inaccurate polarizations.

1. the polarization of write beam. For instance, the polarization is expected to R, but in real experiment, it includes L-polarized component, it will lead to the unexpected correlations of VH.
2. the polarization of read beam. For instance, the polarization is expected to L, but in real experiment, it includes R-polarized component, it will lead to the unexpected correlations of HV.
3. the polarization of write-out and read-out photons, both coincidences of HV and VH will be measured.

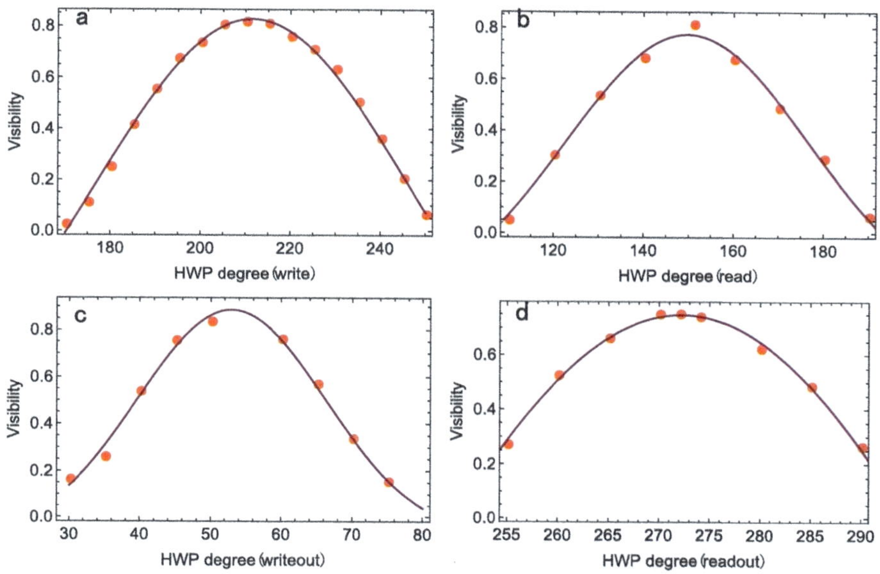

Fig. 5.2 Results of visibility while optimizing the polarizations. a, results for optimizing write beam. **b**, results for optimizing read beam. The read and write beam are optimized several rounds in turn. **c**, results for optimizing write-out photons under eigen bases. **d**, results for optimizing read-out photons under superpositional bases of D/A. During the optimization, the excitation probability is set to 0.03

5.2.1 Photonic Polarization and Bias Magnetic Field

Since the glasscell in the experiment will introduce a relative phase shift between the H/V polarization, the light polarization in the vacuum cannot be directly calibrated in advance. Here we optimize the polarizations based on the measured visibility. As the glasscell doesn't change H/V polarization but lead to a relative phase, the polarizations of write-out and read-out photons can be set to H/V polarization, then the polarizations of write and read beam is optimized to get high visibility. In the experiment, the polarizations of write and read beam are each controlled by a QWP set to 45° and a HWP, we optimize the polarizations by adjusting the HWPs round by round. After optimize the write and read beam, we optimize the polarizations of write-out and read-out photons under superpositional bases ($H \pm V$). The results for the optimization is shown in Fig. 5.2.

Furthermore, because an additional bias magnetic field is required to induce Zeeman splitting (in the east-west direction of the experimental platform), the direction of the magnetic field determines the direction of the quantum magnetic axis and affects the light polarization felt by the atoms. If the magnetic field compensation in the early stage is not sufficient, and there are other directional magnetic fields, it will also affect the entanglement visibility. Therefore, we can further optimize the

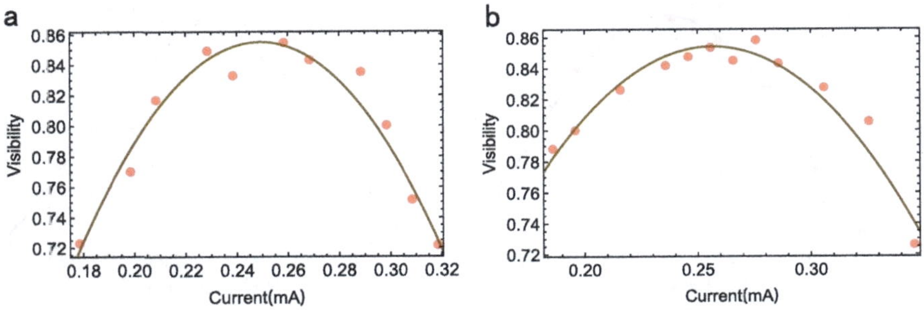

Fig. 5.3 Influence of magnetic field (coil current) perpendicular to the light propagation direction on entanglement visibility. a, up-down direction; **b**, north-south direction

compensation magnetic field by measuring the entanglement visibility under different compensation magnetic field conditions. The results are verified to see if they are consistent with early optimization. In the experiment, the direction of the light is along the east-west direction, so the compensation magnetic field in this direction cannot be optimized here, but it can still be used to calibrate whether the magnetic field along the north-south and up-down directions is well compensated by changing the magnetic coil currents in these two directions. It can be found that the visibility changes with the magnetic field as shown in Fig. 5.3. The magnetic field optimization results here are consistent with the previous compensation, which can be used as an effective verification method for magnetic field compensation.

5.2.2 Accidental Coincidences

As the write process is not a deterministic process, there may be multiple excitations, which will lead to the coincidence of uncorrelated photons. Therefore, experimentally, it is necessary to control the excitation probability to reduce the impact of high-order events. We first measure the excitation probability obtained under different powers of write beam and the visibility measured under the eigen bases and superpositional bases as shown in Fig. 5.4a.

As the power of write beam increases, the excitation probability increases approximately linearly, and the measured visibility decreases approximately linearly. In order to further eliminate the accidental coincidence between the write-out photon and the read-out photon, a new coincidence counting probability p'_{ij} is recalculated as:

$$p'_{ij} = p_{ij} - p_{wi} p_{rj} \tag{5.8}$$

The results for new visibility is shown in Fig. 5.4b. From the results, it can be deduced that after subtracting the accident coincidences, the calculated visibility no longer relies on the excitation probability and stays around 0.96/0.89 under

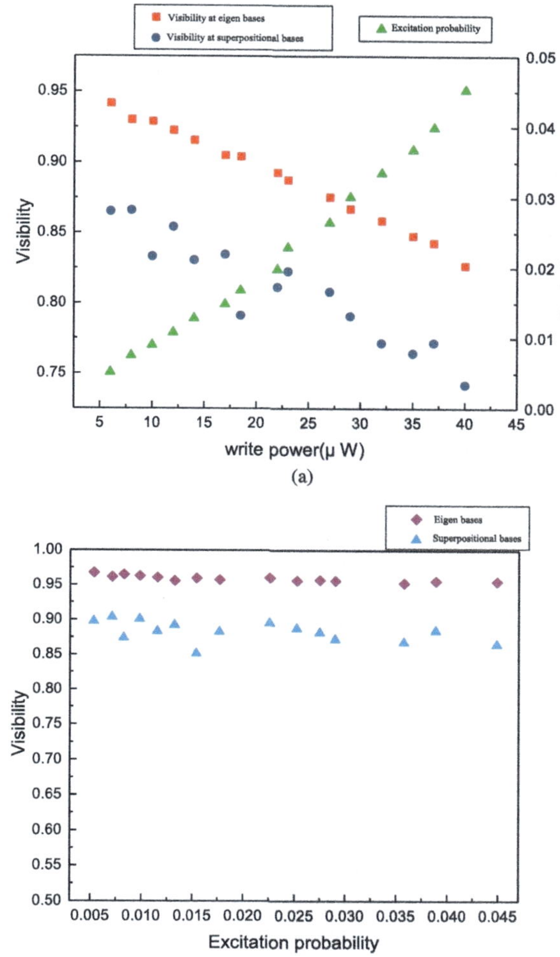

Fig. 5.4 Results for different powers of write beam. a, excitation probability and visibility versus different powers of write beam (~100 ns). **b**, Visibility versus excitation probability after subtracting the accidental coincidences

eigen/superpositional bases. The two values provide an upper bound of the entanglement visibility by controlling the multi-excitation events. It keeps much different from the expected value of 1 under ideal cases, especially under superpositional bases. As seen from the form of the generated entanglement, it is not a maximum entangled state, the coefficient ratio is $\sqrt{3}:\sqrt{2}$, which will lead to the maximum measured visibility under superpositional bases as:

$$V_{DAmax} = \frac{\left(\frac{\sqrt{2}+\sqrt{3}}{\sqrt{3}-\sqrt{2}}\right)^2 - 1}{\left(\frac{\sqrt{2}+\sqrt{3}}{\sqrt{3}-\sqrt{2}}\right)^2 + 1} = 0.98. \quad (5.9)$$

The explanation for the different results under different measurement bases is the interference of write-out channels with different polarizations. While measuring under eigen bases, no interference occurs between the two write-out channels with different polarizations, once the polarization is set accurate, the measured visibility will be high. While measuring under superpositional bases, interference between the two write-out channels with different polarizations should be in consideration, the waveshape, generation probability and frequency, etc. for different polarizations will all influence the interference and thus, the measured visibility. As the different CG coefficients of the transitions during the write process will often result in some internal differences that cannot be resolved by optimizing the external parameters, the best way is to choose atomic energy levels where only the polarizations of write-out channels are different. Hence we change the atomic initial state from $|1, 0\rangle$ to $|1, -1\rangle$, then the CG coefficients are the same for generating different polarized write-out photons, the generated photon-atom entangled state is a maximum entangled state, i.e.,

$$|\Psi_{pa}\rangle = \frac{1}{\sqrt{2}}(|R\downarrow\rangle + |L\uparrow\rangle), \tag{5.10}$$

where $|\downarrow\rangle = |5^2S_{1/2}, F = 2, m_F = -1\rangle$ and $|\uparrow\rangle = |5^2S_{1/2}, F = 2, m_F = +1\rangle$, the phase φ is set to 0. Although the prepared entanglement has advantages over previous state, during the verification process of photon-atom entanglement, the atomic state needs to be converted into photonic state, however, because of the different CG coefficients in the two read-out channels for different polarized read-out photons, the retrieval efficiency and resonant frequency with the ring cavity are different and thus the retrieved photons have intensity difference for different polarizations. While measuring the correlations between write-out and read-out photons under superpositional bases with waveplates and PBS, the measured visibility is less than 0.7, much lower than the expected results. To solve this problem, we consider the photon-atom entangled state is a maximum state after the write process, if we can directly convert the atomic state into the measured superpositional state, then retrieve the superpositional state to read-out photons for measurement, the two read-out channels are no need to interfere with the use of waveplates. It will only influence the photon counting rate but without influencing the visibility. To complete the procedure, we use a Raman operation to realize the transformation of atomic superpositional state.

5.2.3 *Raman Process*

As the two atomic states of generated spinwave $|\uparrow\rangle$ and $|\downarrow\rangle$ is not degenerate under the application of bias magnetic field, with the increasing storage time, the accumulated phase is different for $|\uparrow\rangle$ and $|\downarrow\rangle$, which will lead to a relative phase difference $\varphi(t)$. When the superpositional state $|\downarrow\rangle \pm e^{i\phi}|\uparrow\rangle$ needs to be measured, we can adjust the storage time and let $\varphi(t) + \phi = 0$, then we apply a Raman $\pi/2$ pulse between $|\downarrow\rangle$ and $|\uparrow\rangle$ and convert $|\downarrow\rangle + |\uparrow\rangle$ into $|\downarrow\rangle$, $|\downarrow\rangle - |\uparrow\rangle$ into $|\uparrow\rangle$. Finally a strong read beam is used to map the initial superpositional state into polarization of read-out photons for measurement. The whole process is equivalent to measure the initial atomic state under the basis of $|\downarrow\rangle \pm e^{i\phi}|\uparrow\rangle$. The phase ϕ is controlled by varying the storage time. For example, when the write-out photons is measured at $R \pm L$ while the atomic state is measured at $|\downarrow\rangle \pm |\uparrow\rangle$, Eq. (5.10) can be rewritten as:

$$|\Psi_{pa}\rangle = \frac{1}{2\sqrt{2}}[(R+L)(|\downarrow\rangle + |\uparrow\rangle) + (R-L)(|\downarrow\rangle - |\uparrow\rangle)]. \tag{5.11}$$

For the atomic state, under the operation of Raman $\pi/2$ pulse:

$$|\Psi_{pa}\rangle = \frac{1}{2}[(R+L)|\downarrow\rangle + (R-L)|\uparrow\rangle]. \tag{5.12}$$

For the write-out photon, it is measured by utilizing the waveplates. After applying the read beam, initial state $|\downarrow\rangle$ and $|\uparrow\rangle$ can be converted into polarization state of read-out photon and then be measured to get the results of measurement under superpositional bases. Going further, we can realize any measurements under superpositional bases $\alpha|\downarrow\rangle \pm \beta e^{i\phi}|\uparrow\rangle$ by adjusting the pulse width of Raman beam.

In our experiment, as the frequency difference between the two atomic states is less than 1 MHz, we use two laser beams to couple the two states, the laser frequency of the two beam is almost the same but with opposite polarizations of σ^+ and σ^-. It locates in the middle of $|F = 2\rangle$ and $|F' = 1, 2\rangle$ and is 407 MHz detuned from both of the two excited states, which can help the Raman process make use of both excited states. As the sign of CG coefficients is opposite when the Raman process occurs via different excited states, it can lead to a enhancement to the Rabi frequency. In principle, if the two Raman beams come from two independent lasers, the relative phase should be locked to ensure that the phase will not be disturbed during the atomic state flipping process and will not affect the fidelity of the Raman operation. Here, thanks to the small bias magnetic field, the frequency difference is small, we can use a H polarized laser to approximately replace the two independent Raman beams required. The single H polarized Raman beam has already included the σ^+ and σ^- polarized beam which are the Raman beams discussed above. As they come from the same laser, its relative phase keeps stable, no extra phase-locking device is needed. In the experimental setup, it is overlapped with the depumper laser via a PBS, and close to the write beam direction. The beam waist radius is set to 1.8 mm, which is nearly parallel light.

Figure 5.5 shows the measured retrieval efficiency of different polarized read-out photons conditioned on the detection of write-out photons with different polarizations. It oscillates because the states transfer between the two atomic spin states $|\downarrow\rangle$ and $|\uparrow\rangle$, i.e., Rabi oscillations between $|\downarrow\rangle$ and $|\uparrow\rangle$. It can be deduced from the measured curve that the average efficiency of Raman $\pi/2$ pulse for flipping the atomic state is around 0.97. The corresponding Raman pulse width can be in principle determined through the oscillation curve. In real experiments, although the pulse is expected to have a uniform intensity, considering the pulse is modulated through

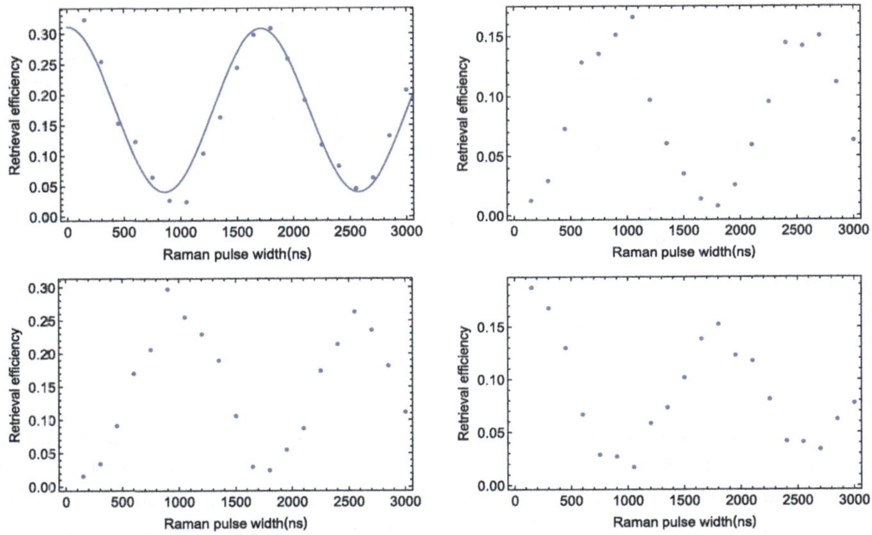

Fig. 5.5 Rabi oscillation characterized by correlations of write-out and read-out photons with four polarization combinations. **a**, write-out: R, read-out: L; **b**, write-out: R, read-out: R; **c**, write-out: L, read-out: R; **d**, write-out: L, read-out: R

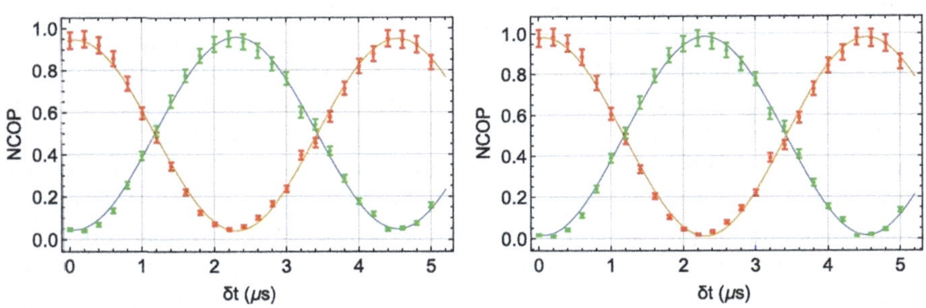

(a) (b)

Fig. 5.6 Normalized coincidence probabilities (NCOP) as a function of Raman pulse delay δt relative to the detection of write-out photons. The difference between **a** and **b** is the raw measured results (**a**) and results after subtracting the accidental coincidences between write-out and read-out photons (**b**)

AOMs, it will inevitably cause variational laser intensity at the rising and falling edge. To make the pulse width of Raman $\pi/2$ pulse more accurate, we often take the entanglement visibility as referenced signal and optimize the width in a small range near the deduced width from oscillation curve. Through the Raman operation, we measure the entanglement visibility under superpositional bases versus the Raman pulse delay δt relative to the write pulse, as shown in Fig. 5.6. The horizontal axis is the time delay of the Raman pulse from the detection time of write-out photons,

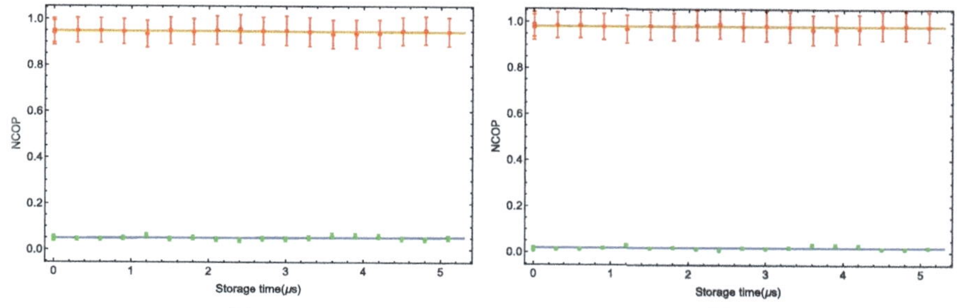

Fig. 5.7 Normalized coincidence probabilities (NCOP) of correlated/uncorrelated photons ($RL + LR$, $RR + LL$) as a function of storage time under eigen basis. a, the raw measured results. **b**, the results after subtracting the accidental coincidences between write-out and read-out photons

and the vertical axis is the normalized coincidence probability of the write-out and read-out photons that should be polarization correlated/uncorrelated at the beginning of generating photon-entanglement. It corresponds to the coincident probability of $(R + L)|\downarrow\rangle$ and $(R - L)|\uparrow\rangle$, $(R - L)|\downarrow\rangle$ and $(R + L)|\uparrow\rangle$.

Figure 5.6b is the result after subtracting the accidental coincidences between write-out and read-out photons, the visibility is improved from 0.91 to 0.97 after excluding the influence of accidental coincidence, it can be seen that this factor is an important factor affecting the current visibility. Although it can be solved by reducing the excitation probability, it will also reduce the final counting rate of the experiment. Therefore, after weighing the brightness and entanglement visibility of the multi-photon coincidence in the experiment, we will not reduce the excitation probability to obtain a better visibility. This effect can be solved by using Rydberg atoms to produce deterministic photon and atomic entanglement [9, 10].

Going further, in order to measure the fidelity of the prepared photon-atom entanglement source, we carefully measured the correlated and uncorrelated photons under the eigen bases versus the storage time, as shown in Fig. 5.7, it can be calculated the visibility under the eigen basis is about 0.91, after deducting the accidental coincidence, the visibility is 0.96. It can be seen that the results under the eigen basis and the superposition basis are almost the same now. Thus, great achievements have been made in optimizing the visibility of photon-atom entanglement. The main influence factor now lies in the impact of the excitation probability on the accidental coincidences. Here we will not continue to optimize cause we have known the influence factors.

5.2.4 Multi-excitation Events

Now we analyze how the multi-excitation events affect the measured visibility, for simplicity, we consider the internal probability of p to generate one write-out photon and p^2 to generate two two write-out photons. The overall transmission and detection efficiency for a write-out photon is p_{dw}. Then the state between write-out and read-out photons can be written as:

$$|\Psi\rangle = \sqrt{p}(|L_w R_r\rangle + |R_w L_r\rangle) + p(|L_w L_w R_r R_r\rangle + |R_w R_w L_r L_r\rangle) + p|L_w R_w R_r L_r\rangle, \quad (5.13)$$

where the subscript w, r represent the write-out and read-out photons, respectively. Assuming the intrinsic retrieval efficiency is η, and overall transmission and detection efficiency of read-out photons is η_{dr}. Then we can calculate the contributed probability of detecting $L_w R_r$, $L_w L_r$, $R_w R_r$ and $R_w L_r$ of each term as following: Items in the first bracket only contributes to LR or RL:
$P_1 = 2pp_{dw}\eta p_{dr}$.
Items in the second bracket only contributes to LR or RL:
$P_2 = 2p^2[1-(1-p_{dw})^2][\eta^2(1-(1-p_{dr})^2)+\eta(1-\eta)p_{dr}]$.
Items in the last bracket contributes to LR or RL:
$P_3 = 2p^2[p_{dw}(1-p_{dw})\eta p_{dr}(1-\eta p_{dr})]$.
Items in the last bracket contributes to LL or RR:
$P_4 = 2p^2[p_{dw}(1-p_{dw})\eta p_{dr}(1-\eta p_{dr})]$.

Here we have excluded the events that the write-out(read-out) photons with opposite polarizations are simultaneously detected as these events can be technically distinguished. As the measured visibility is calculated through the coincident counts of $L_w R_r$, $L_w L_r$, $R_w R_r$ and $R_w L_r$, we can calculate the visibility as:

$$V = \frac{P_1 + P_2 + P_3 - P_4}{P_1 + P_2 + P_3 + P_4}. \quad (5.14)$$

When considering the specific parameters in the experiment, the visibility is 0.97, an upper bound at current experimental parameters.

5.2.5 Initial State Purity

As the efficiency of pumping all atoms into state $|1,-1\rangle$ can not reach 100%, some atoms can still locate in state $|m_F = 0, 1\rangle$. Assuming the probability of atoms in the three states ($|F = 1, m_F = -1, 0, 1\rangle$) is p_1, p_2, p_3, respectively. After the write process, the induced states between write-out and atomic states corresponding to the three initial states are:

$$|1,-1\rangle:\sigma^+|-1\rangle_{-1}\cdot\left(-\sqrt{\frac{1}{12}}\right)\cdot\sqrt{\frac{1}{4}}-\sigma^-|1\rangle_{-1}\cdot\left(-\sqrt{\frac{1}{12}}\right)\cdot\sqrt{\frac{1}{4}},\tag{5.15}$$

$$|1,0\rangle\ :\sigma^+|0\rangle_0\cdot\left(-\sqrt{\frac{1}{4}}\right)\cdot\sqrt{\frac{1}{4}}-\sigma^-|2\rangle_0\cdot\left(-\sqrt{\frac{1}{4}}\right)\cdot\sqrt{\frac{1}{6}},\tag{5.16}$$

$$|1,1\rangle\ :\sigma^+|1\rangle_1\cdot\left(-\sqrt{\frac{1}{2}}\right)\cdot\sqrt{\frac{1}{6}},\tag{5.17}$$

where $\sigma^{\pm}$ is equivalent to R/L, the subscripts represent the m_F of initial atomic states in $|F=1\rangle$, $|i\rangle(i=-2,-1,0,1,2)$ represents the $|m_F\rangle$ of atomic states in $|F=2\rangle$. After the read process, state between write-out and read-out photons can be written as:

$$|1,-1\rangle:|\psi_1\rangle=-\sigma^+\sigma^-e^{-i2\Delta t}\cdot\sqrt{\frac{1}{48}}+\sigma^-\sigma^+\cdot\sqrt{\frac{1}{48}},\tag{5.18}$$

$$|1,0\rangle\ :|\psi_2\rangle=-\sigma^+\sigma^-\sqrt{\frac{1}{16}}+\sigma^-\sigma^+e^{i2\Delta t}\cdot\sqrt{\frac{1}{24}},\tag{5.19}$$

$$|1,1\rangle\ :|\psi_3\rangle=-\sigma^+\sigma^-e^{i2\Delta t}\cdot\sqrt{\frac{1}{12}},\tag{5.20}$$

where Δ is the Zeeman splitting of energy level. The final density matrix of the state between write-out and read-out photons is:

$$\rho=p_1|\psi_1\rangle\langle\psi_1|+p_2|\psi_2\rangle\langle\psi_2|+p_3|\psi_3\rangle\langle\psi_3|.\tag{5.21}$$

When the entanglement visibility is measured at circular polarized bases, R, L represent σ^+,σ^-, we define the bases LR, LL, RR, RL, as $(1,0,0,0)^T$, $(0,1,0,0)^T$, $(0,0,1,0)^T$ and $(0,0,0,1)^T$, respectively, which correspond to the elements of the density matrix. The visibility is expected to be:

$$V_{RL}=\frac{\rho_{11}+\rho_{44}-\rho_{22}-\rho_{33}}{\rho_{11}+\rho_{44}+\rho_{22}+\rho_{33}}.\tag{5.22}$$

As $\rho_{22}+\rho_{33}=0$, it means the visibility can reach 100% under eigen basis. If it is measured under superpositional basis, for instance, H and V, then:

$$\mathrm{RR}=\frac{1}{2}\begin{pmatrix}1\\ i\\ i\\ -1\end{pmatrix};\mathrm{RL}=\frac{1}{2}\begin{pmatrix}1\\ -i\\ i\\ 1\end{pmatrix};\mathrm{LR}=\frac{1}{2}\begin{pmatrix}1\\ i\\ -i\\ 1\end{pmatrix};\mathrm{LL}=\frac{1}{2}\begin{pmatrix}1\\ -i\\ -i\\ -1\end{pmatrix}.$$

Now the visibility can be calculated as:

$$V_{HV} = \frac{C_{HV} + C_{VH} - C_{HH} - C_{VV}}{C_{HV} + C_{VH} + C_{HH} + C_{VV}}, \tag{5.23}$$

where C corresponds to the four diagonal elements of the density matrix.

5.2.6 Photon-Atom Entanglement Fidelity

For a maximum entanglement like $|\Psi\rangle = 1/\sqrt{2}(|RL\rangle + |LR\rangle)$, the density matrix can be written as:

$$\begin{aligned} \rho =& |\Psi\rangle\langle\Psi| \\ =& \frac{1}{4}[(|RL\rangle\langle RL| + |LR\rangle\langle LR| - |RR\rangle\langle RR| - |LL\rangle\langle LL|) \\ &+ (|HH\rangle\langle HH| + |VV\rangle\langle VV| - |HV\rangle\langle HV| - |VH\rangle\langle VH|) \\ &+ (|DD\rangle\langle DD| + |AA\rangle\langle AA| - |DA\rangle\langle DA| - |AD\rangle\langle AD|) + I]. \end{aligned} \tag{5.24}$$

By measuring the visibility under eigen basis and two superpositional bases, the entanglement fidelity can be calculated as:

$$F = \frac{1}{4}(V_{RL} + V_{HV} + V_{DA} + 1). \tag{5.25}$$

For the experimentally measured data, we can estimate that the fidelity of the entangled state compared to the expected maximum entangled state is about 0.933. After deducting the accidental coincidences, it is 0.975, which is far beyond the classical threshold of 0.5, and proves the high quality of the entanglement source. It can be used to carry out the entanglement of multiple photons and multiple atomic ensembles.

5.2.7 Efficiency of Photon-Atom Entanglement

An important indicator to measure the efficiency of photon-atom entanglement is p, which includes the efficiency of entanglement generation (write process, excitation probability p_w) and the efficiency of entanglement verification (read process, measured retrieval efficiency p_r), $p = p_w p_r$. In the previous experiment of atom-atom entanglement using photon-atom entanglement source [11], in each experimental trial, p is only 0.0004. Under this condition, if experiment of three quantum memories is carried out, the overall efficiency will be only on the order of 10^{-11}, which is very difficult to proceed. In this thesis, we increase the intrinsic retrieval efficiency to about 0.9 by enhancing the interaction between photons and atoms through the ring cavity, and set the detuning of write and read beam from the excited state with opposite signs, eliminating the needs to use etalon and atomic filtering cell in previous experiments. By doing this, it greatly improves the single photon transmission efficiency, so that the maximum measured retrieval efficiency is about $\eta_r = 0.4$. Since the signal-to-noise ratio in the experiment is determined by the ratio, that is, the second-order correlation function η_r/p_w, the higher the measured retrieval efficiency can be obtained in the experiment means that we can use a higher excitation probability during the write process. Therefore, we are able to increase the experimental parameter p to about 0.006. Under this condition, the efficiency of the entanglement experiment containing three atomic ensembles will be at least three orders of magnitude higher than before. A further improvement can be also achieved through the Rydberg blockade to deterministically generate write-out photons and photon-atom entanglement [9, 10, 12] to realize the entanglement among multiple cold atomic ensembles.

5.3 Entanglement Between Two Cold Atomic Ensembles

After obtaining a high-fidelity and efficiently retrieved photon-atom entanglement source, we begin to connect the two cold atomic ensembles by using two pairs of photon-atom entanglement generated by quantum memories and interfere the two photons to achieve entanglement swapping and establish entanglement between the remaining two cold atomic ensembles [11].

Figure 5.8 shows the optical path for interference of two write-out photons and operation of entanglement swapping [13]. The write-out photons is released via a

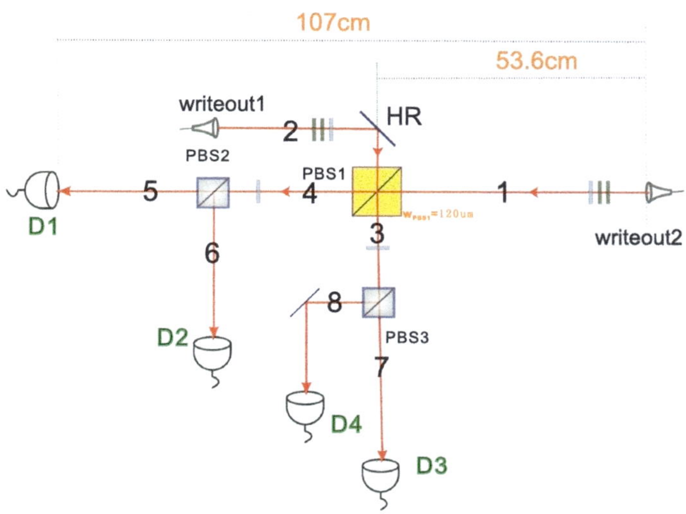

Fig. 5.8 Experiment setup of Bell state measurement for generating entanglement between two cold atomic ensembles

fiber collimator (SK company, A_{11}) from a single-mode fiber with mode diameter about 5 μm. By adjusting the fiber collimators, the two write-out photons reach its beam waist of 120 μm at a distance of 53.6 cm from the related collimators. A PBS (PBS1) is placed here to combine the two photons and realize their interference.

For the two pair of photon-atom entanglement:

$$\begin{aligned} |\Psi_1\rangle &= \frac{1}{\sqrt{2}}(|R\downarrow\rangle + |L\uparrow\rangle), \\ |\Psi_2\rangle &= \frac{1}{\sqrt{2}}(|R\downarrow\rangle + |L\uparrow\rangle). \end{aligned} \tag{5.26}$$

For simplicity, we use two QWPs and one HWP to perform a unitary operation on the photonic polarizations: $R \rightarrow H, H \rightarrow H + V$. Thus the two photon-entanglement can be rewritten as:

$$\begin{aligned} |\Psi\rangle_1 &= \frac{1}{\sqrt{2}}(|H\downarrow\rangle + |V\uparrow\rangle), \\ |\Psi\rangle_2 &= \frac{1}{\sqrt{2}}(|H\downarrow\rangle + |V\uparrow\rangle). \end{aligned} \tag{5.27}$$

After passing through PBS1, HWPs are used to transform the H/V basis to D/A basis, the evolution process is described as following:

$$\begin{aligned} |\Psi\rangle_1|\Psi\rangle_2 &= \frac{1}{2}(|H_1H_2\downarrow\downarrow\rangle + |V_1V_2\uparrow\uparrow\rangle + |H_1V_2\downarrow\uparrow\rangle + |V_1H_2\uparrow\downarrow\rangle) \\ &\xrightarrow{\textbf{PBS1}} \frac{1}{2}(|H_3H_4\downarrow\downarrow\rangle + |V_4V_3\uparrow\uparrow\rangle + |H_3V_3\downarrow\uparrow\rangle + |V_4H_4\uparrow\downarrow\rangle) \\ &\rightarrow \frac{1}{\sqrt{2}}(|H_3H_4\downarrow\downarrow\rangle + |V_4V_3\uparrow\uparrow\rangle)\text{(post selecet photons not in the same path)} \\ &\xrightarrow{\textbf{HWP}} \frac{1}{\sqrt{2}}(|D_3D_4\downarrow\downarrow\rangle + |A_4A_3\uparrow\uparrow\rangle) \\ &= \frac{1}{2\sqrt{2}}(|(H_3+V_3)(H_4+V_4)\downarrow\downarrow\rangle + |(H_4-V_4)(H_3-V_3)\uparrow\uparrow\rangle) \\ &\xrightarrow{\textbf{PBS2,3}} \frac{1}{2\sqrt{2}}((H_7H_5+V_6V_8)(|\downarrow\downarrow\rangle + |\uparrow\uparrow\rangle) + (H_7V_6 - H_5V_8)(|\downarrow\downarrow\rangle - |\uparrow\uparrow\rangle)) \end{aligned} \tag{5.28}$$

Here the subscripts represent the optical path shown in Fig. 5.8. By measuring the two photons, the coincidences will tell which entangled state the atomic ensembles will collapse to. If coincidences between detector D1, D3 or D2, D4 are measured, the corresponding entangled state between the two cold atomic ensembles is:

$$|\Psi\rangle_{aa1} = \frac{1}{\sqrt{2}}(|\downarrow\downarrow\rangle + |\uparrow\uparrow\rangle). \tag{5.29}$$

When coincidences between detector D1, D4 or D2, D3 are measured, the corresponding entangled state between the two cold atomic ensembles is:

$$|\Psi\rangle_{aa2} = \frac{1}{\sqrt{2}}(|\downarrow\downarrow\rangle - |\uparrow\uparrow\rangle). \tag{5.30}$$

Considering the evolution of $|\downarrow\rangle$ and $|\uparrow\rangle$ with increasing storage time, if we have known the atom-atom state is $|\Psi\rangle_{aa2}$, we can always let the state evolve to $|\Psi\rangle_{aa1}$ by controlling the storage time. By doing this, we can double the brightness of generation and verification of entangled state $|\Psi\rangle_{aa1}$.

5.3.1 Pulse Shape of Write Beam

In order to achieve high quality atom-atom entanglement, the key is that the write-out photons from the two quantum memories need to be perfectly interfered during the entanglement swapping process. This requires that the indistinguishability of the two write-out photons is sufficiently high. As the write-out photon mainly comes from the spontaneous Raman scattering of the write beam, the first step is to adjust the pulse shape of the write beam. We first analyze the factors that affect the write beam in the experiment. For a real experimental system, in order to keep the shape of the write-out photons from the two ensembles consistent, the key point is to adjust the shape of the two write pulses to be the same. This can be achieved by adjusting the optical path for generating write beam except that the radio frequency signal applied to AOM involved in the application of switch write beam needs to be optimized separately.

In the experiment, the write beam is usually generated by a TTL gate signal acting on a switch that controls the radio frequency source and then driving the AOM, so that the radio frequency signal is output under the action of a switch, thereby turning on the write beam. In the previous experiment, the RF signal is always output, and then outputs to the AOM through a switch controlled by a gate signal, as shown in Fig. 5.9. However, there exists a problem. When the gate signal comes, the phase of the RF signal arriving at the 'switch-on' position may be not the same, for instance, 0 (a) and $\pi/2$ (c), which will result in different outputs of the RF. Since the phase of the output signal after the RF switch is turned on always starts from 0 (a), it cannot be directly output from the phase $\pi/2$ (c). When the gate signal comes, if the RF signal is exactly at phase 0, then it can output a complete RF signal through the switch (b), if the RF signal is exactly at phase $\pi/2$, the output signal will quickly start from 0 in the first half cycle and then quickly return to 0, connect the signal of the next cycle (d). It is similar when the gate signal is closed.

Since the pulse width used in our experiment is very short (50 ns), the required RF signal usually has only 5 cycles. If one cycle changes, the Fourier transform during the whole process will cause the RF signal to contain multiple frequency components. It will not only affect the frequency of write beam, but also affect the pulse shape. As shown in Fig. 5.10a, it is the pulse shape of the write beam, the shaded part is where the write light may appear, which has an uncertain time about 10 ns, while the entire width is only 50 ns, which will affect the indistinguishability

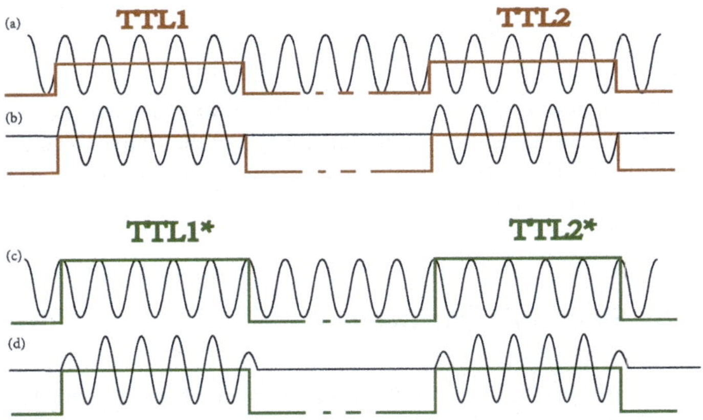

Fig. 5.9 Radio frequency signal used to control the switch of write beam

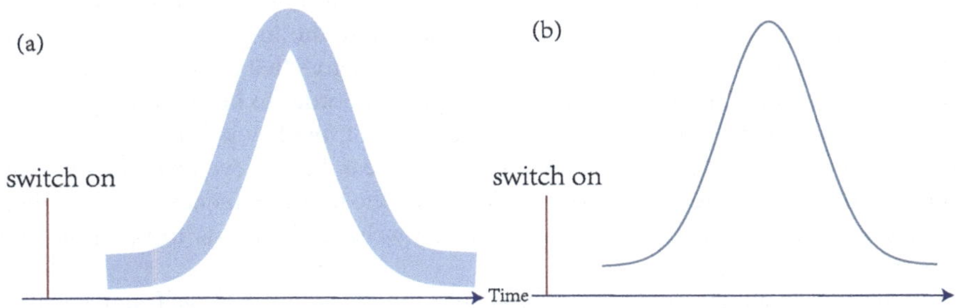

Fig. 5.10 The pulse shape of write beam before and after optimization of the radio frequency signal

of the write-out photons. To solve this problem, we use a special signal source. When the gate signal arrives, the radio frequency source will always output 5 complete sine waves from the phase of 0 to the AOM. The write beam obtained at this time has basically no deviation in time relative to the gate signal, the generated pulse is shown in Fig. 5.10b.[1] Furthermore, we measure the changes of pulse of write beam when outputting with different initial phases as shown in Fig. 5.11. It can be seen that the pulse intensity is the lowest when output from $\pi/2$.

[1] Since I did not take a screenshot of the oscilloscope in time, the diagram is roughly drawn based on the oscilloscope image, but the phenomenon described above is true and can be observed with an oscilloscope.

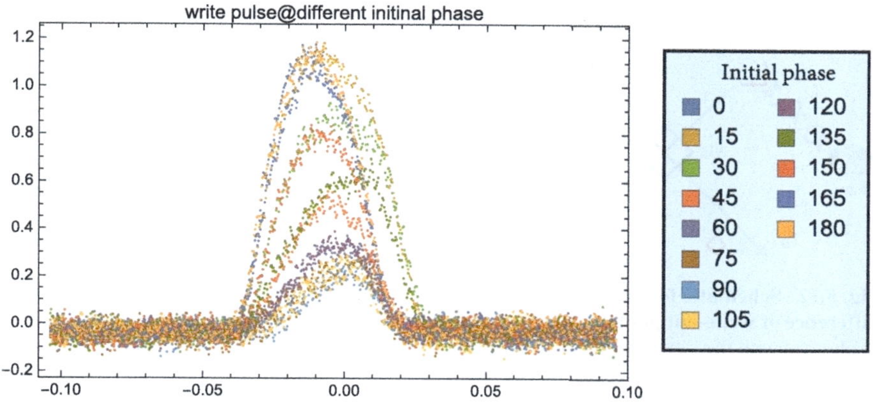

Fig. 5.11 The pulse shape of write beam versus different initial phases of RF. When measuring, we first output a DC voltage corresponding to the amplitude of the initial phase value, and then continuously output a complete sine signal from here

5.3.2 Indistinguishability of Write-Out Photons

After optimizing the write beam, we begin to consider the difference of write-out photons caused by other reasons. In fact, due to the energy level splitting caused by Zeeman splitting under bias magnetic field or fluctuated magnetic field, there will be some differences in the frequency of photons with different polarizations.

First, we analyze the influence of the different frequencies and the shape of the photon wave packet due to the splitting of the energy levels. We mark the frequency of write-out photons with σ^+ polarization as ω_+ and frequency of photons with σ^- polarization as ω_-. Photons with ω_- come from the transition of clock states, which is insensitive to magnetic field. As shown in Fig. 5.12a, entanglement between write-out photons (locating in path 1 and 2) and atomic states (A and B) is:

$$\begin{aligned}|\Psi\rangle_{1A} &= \frac{1}{\sqrt{2}}(|H,\omega_+\rangle_1|\downarrow\rangle_A + |V,\omega_-\rangle_1|\uparrow\rangle_A),\\ |\Psi\rangle_{2B} &= \frac{1}{\sqrt{2}}(|H,\omega_+\rangle_2|\downarrow\rangle_B + |V,\omega_-\rangle_2|\uparrow\rangle_B).\end{aligned} \tag{5.31}$$

According to the previous sections, photons going into different outputs 3, 4 are the events we consider. Assuming the detection time of photons in 3 and 4 is t_3 and t_4, the state between photons and atoms after the interference PBS is described as:

$$|\Psi\rangle_a = \frac{1}{\sqrt{2}}[|H,\omega_+\rangle_4|H,\omega_+\rangle_3|\downarrow\rangle_A|\downarrow\rangle_B + |V,\omega_-\rangle_3|V,\omega_-\rangle_4|\uparrow\rangle_A|\uparrow\rangle_B]. \tag{5.32}$$

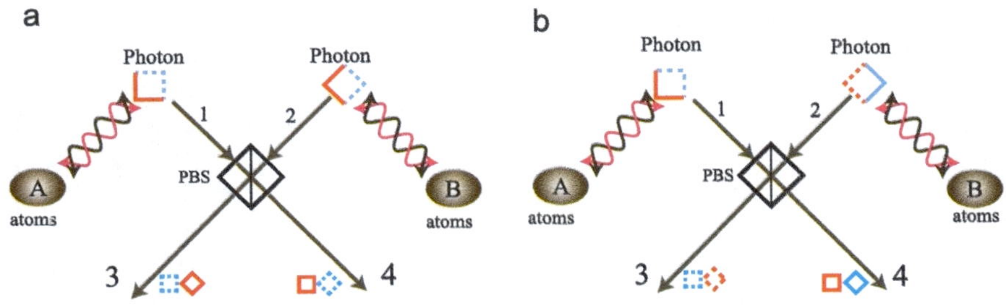

Fig. 5.12 Schematic for generating atom-atom entanglement in consideration of frequency difference of write-out photons. a, photons with the same frequency have the same polarizations. **b**, photons with the same frequency has the opposite polarizations. Different frequency is marked by solid line (ω_+) and doted line (ω_-), different polarizations are marked as red(H) and blue(V)

The photons from the two quantum memories and with two polarizations are Fourier transformed from the frequency form to the time form and can be written as:

$$\begin{aligned}|\omega_+\rangle_3 &= \int dte^{-i\omega_+ t} f_3(t)|t\rangle,\\ |\omega_-\rangle_3 &= \int dte^{-i\omega_- t} g_3(t)|t\rangle,\\ |\omega_+\rangle_4 &= \int dte^{-i\omega_+ t} f_4(t)|t\rangle,\\ |\omega_-\rangle_4 &= \int dte^{-i\omega_- t} g_4(t)|t\rangle,\end{aligned} \tag{5.33}$$

where f and g represent the distribution function of the photons in the time domain. Substitute them into Eq. 5.32, we can get:

$$|\Psi\rangle_a = \frac{1}{2}\int\int dt_3 dt_4 e^{-i\omega_+(t_3+t_4)} f(t_3) f(t_4)[|\Psi\rangle^+_{AB} \otimes |\Phi\rangle^+ + |\Psi\rangle^-_{AB} \otimes |\Phi\rangle^-]. \tag{5.34}$$

Here,

$$|\Phi\rangle^{\pm} = \frac{1}{\sqrt{2}}(|H, t_4\rangle_4 |H, t_3\rangle_3 \pm |V, t_3\rangle_3 |V, t_4\rangle_4). \tag{5.35}$$

$|\Phi\rangle^{\pm}$ can be further distinguished by measuring the photons in 3, 4 under D/A basis, thus the Bell state measurement, the atom-atom state can be collapsed into an entangled state:

$$|\Psi\rangle^{\pm}_{AB} = |\downarrow\rangle_A |\downarrow\rangle_B \pm e^{-i(\omega_- - \omega_+)(t_3+t_4)} \frac{g_3(t_3) g_4(t_4)}{f_3(t_3) f_4(t_4)} |\uparrow\rangle_A |\uparrow\rangle_B. \tag{5.36}$$

For a common consideration where

$$\begin{aligned}
&\omega_+ = \omega_-, \\
&f_3 = f_4 = g_3 = g_4, \\
&e^{-i(\omega_- - \omega_+)(t_3+t_4)} \frac{g_3(t_3)g_4(t_4)}{f_3(t_3)f_4(t_4)} = 1,
\end{aligned} \tag{5.37}$$

we can always get the atom-atom entanglement as the maximum entangled state after the entanglement swapping. Back to the real parameters, the generated atom-atom entanglement is a detection-time depended state. Taking into account the different distribution of detection time under repeated experiments, we will eventually get a mixed state [14], the quality of generated atom-atom entangled state will also be influenced. Before we optimize this influence, if we measure the visibility under eigen basis, the result is about 0.8 and independent of the interference, while we measure the visibility under superpositional basis, the result is about 0.65, obviously lower than expected.

From Eq. (5.36), we can know that when the four distribution function is the same, the main influence factors of the uncertain phase:

$$\phi = (\omega_- - \omega_+)(\delta t_3 + \delta t_4), \tag{5.38}$$

where δt_3, δt_4 are the uncertainty of photon detection time, it usually means the time width of detected photons. It can be seen that it is very helpful to use a short pulse of write beam and a small bias magnetic field. The final measured visibility under superpositional basis is:

$$V = \frac{\left|\frac{1+e^{-i\phi}}{1-e^{-i\phi}}\right|^2 - 1}{\left|\frac{1+e^{-i\phi}}{1-e^{-i\phi}}\right|^2 + 1}. \tag{5.39}$$

The visibility that can be obtained by simulating different pulse widths and different magnetic field fields is shown in Fig. 5.13, to acquire high quality atom-atom entanglement, we experimentally set the magnetic field to 140 mG and the full pulse width of write beam to 50 ns.

In order to eliminate the influence of frequency difference between different polarized photons, we transform the polarization of photons before interference, in detail, we flip the photon polarization in path 2 (Fig. 5.12b), i.e., the polarization transformation now is: $R \rightarrow V, H \rightarrow H + V$, then the photon-atom entanglement $|\Psi\rangle_{2B}$ is:

$$|\Psi\rangle_{2B} = \frac{1}{\sqrt{2}}(|V, \omega_+\rangle_2|\downarrow\rangle_B + |H, \omega_-\rangle_2|\uparrow\rangle_B. \tag{5.40}$$

Similarly, we only consider cases that two photons go to different output ports of PBS:

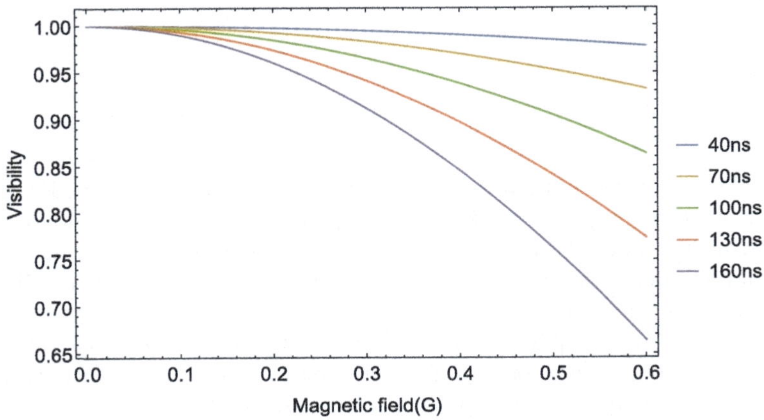

Fig. 5.13 Visibility under different bias magnetic field and different time distributions ($\delta t_3 + \delta t_4$) of detected photons

$$
\begin{aligned}
|\Psi\rangle_b =& \frac{1}{\sqrt{2}}(|H, \omega_+\rangle_4 |H, \omega_-\rangle_3 |\downarrow\rangle_A |\uparrow\rangle_B + |V, \omega_-\rangle_3 |V, \omega_+\rangle_4 |\uparrow\rangle_A |\downarrow\rangle_B) \\
=& \frac{1}{\sqrt{2}} |\omega_+\rangle_4 |\omega_-\rangle_3 (|H\rangle_4 |H\rangle_3 |\downarrow\rangle_A |\uparrow\rangle_B + |V\rangle_3 |V\rangle_4 |\uparrow\rangle_A |\downarrow\rangle_B) \\
=& \frac{1}{2\sqrt{2}} |\omega_+\rangle_4 |\omega_-\rangle_3 [(|\downarrow\rangle_A |\uparrow\rangle_B + |\uparrow\rangle_A |\downarrow\rangle_B) \otimes (|H\rangle_4 |H\rangle_3 + |V\rangle_3 |V\rangle_4) \\
& + (|\downarrow\rangle_A |\uparrow\rangle_B - |\uparrow\rangle_A |\downarrow\rangle_B) \otimes (|H\rangle_4 |H\rangle_3 - |V\rangle_3 |V\rangle_4)].
\end{aligned}
\tag{5.41}
$$

It can be seen that through this simple operation, photons with the same frequency interfere and enter into the same output port of PBS, the interference between frequency distinguished photons is avoided. It not only eliminates the influence of frequency differences in the experiment on the quality of entanglement, but also eliminates the influence caused by the difference in the shape of the photon wave packet with different frequency at different polarizations. Through further projection measurement, we can obtain the maximum entangled state for the two cold atomic ensembles. By performing the optimization, the fidelity of the atom-atom entanglement measured in the experiment can be improved from 0.81 to 0.85.

5.3.3 Imperfection of Polarization

Next, we consider whether the effect on the entanglement visibility can be ignored if the accuracy is not enough during the polarization rotation process. For example, the transformation of polarization of one write-out photon before interference is not accurate enough, and the resulting entangled state is written as:

$$|\Psi_1\rangle = \frac{1}{\sqrt{2}}(|cos\theta H + sin\theta V, \downarrow\rangle + |cos\theta V - sin\theta V, \uparrow\rangle). \quad (5.42)$$

The evolution of the two pairs of photon-atom entanglement can now be described as:

$$\begin{aligned}\Psi &= (H|\downarrow\rangle + V|\uparrow\rangle) \otimes [(cos\theta H + sin\theta V)|\downarrow\rangle + (cos\theta V - sin\theta H)|\uparrow\rangle] \\ &\Rightarrow cos\theta[HH|\downarrow\rangle|\downarrow\rangle + VV|\uparrow\rangle|\uparrow\rangle] - sin\theta[HH|\downarrow\rangle|\uparrow\rangle - VV|\uparrow\rangle|\downarrow\rangle] \\ &= cos\theta[DD(|\downarrow\rangle|\downarrow\rangle + |\uparrow\rangle|\uparrow\rangle) + AA(|\downarrow\rangle|\downarrow\rangle + |\uparrow\rangle|\uparrow\rangle) + DA(|\downarrow\rangle|\downarrow\rangle - |\uparrow\rangle|\uparrow\rangle) \\ &+ AD(|\downarrow\rangle|\downarrow\rangle - |\uparrow\rangle|\uparrow\rangle)] - sin\theta[DD(|\downarrow\rangle|\uparrow\rangle - |\uparrow\rangle|\downarrow\rangle) + DA(|\downarrow\rangle|\uparrow\rangle + |\uparrow\rangle|\downarrow\rangle) \\ &+ AD(|\downarrow\rangle|\uparrow\rangle + |\uparrow\rangle|\downarrow\rangle) - AA(|\downarrow\rangle|\uparrow\rangle - |\uparrow\rangle|\downarrow\rangle)].\end{aligned} \quad (5.43)$$

The final visibility is:

$$V = cos^2\theta - \sin^2\theta. \quad (5.44)$$

Since the polarizations are adjusted with waveplates by rotating the angle, and the accuracy of the angle when we adjust is less than 1°, its influence on the visibility is less than 1%.

5.3.4 From Photon-Atom Entanglement to Atom-Atom Entanglement

When light is entangled with atoms, due to specific experimental limitations, the fidelity obtained is limited, assuming the signal to noise ratio is 1: δ^2 (the corresponding visibility is $\frac{1-\delta^2}{1+\delta^2}$), then the photon-atom entanglement can be written as:

$$|\Psi\rangle_{ap} = \frac{1}{\sqrt{2}}(|H\downarrow\rangle + |V\uparrow\rangle + \delta|H\uparrow\rangle + \delta|V\downarrow\rangle). \quad (5.45)$$

The acquired atom-atom state is:

$$\begin{aligned}|\Psi\rangle_{ap} \otimes |\Psi\rangle_{ap} \Rightarrow &|HH\downarrow\downarrow\rangle + |VV\uparrow\uparrow\rangle + \delta|HH\downarrow\uparrow\rangle \\ &+ \delta|VV\uparrow\downarrow\rangle + \delta|VV\downarrow\uparrow\rangle + \delta|HH\uparrow\downarrow\rangle.\end{aligned} \quad (5.46)$$

In comparison with the ideal entangled state $|\Psi\rangle_{aa}^{ideal} = |\downarrow\downarrow\rangle + |\uparrow\uparrow\rangle$, the measured SNR would be $1 : 2\delta^2$, reduced by half compared to the SNR of single pair of photon-atom entanglement. The visibility becomes $\frac{1-2\delta^2}{1+2\delta^2}$. As a result, if the measured SNR of single pair of photon-atom entanglement is 20: 1, then the expected visibility for atom-atom entanglement should be around 10: 1, which is consistent with the measured result of 9: 1 during the experiment.

5.3.5 Experimental Results

After optimizing the aforementioned various experimental parameters and analyzing some of the factors that affect atom-atom entanglement, we begin to measure the entanglement visibility of atom-atom entanglement. The measured visibility under eigen basis is about 0.82. When measuring under the superpositional basis, we still adjust the storage time and use the Raman $\pi/2$ pulse to manipulate the atomic state, and then retrieve it out for detection. The measured visibility is about 0.78. Finally, The estimated fidelity of entanglement between two cold atomic ensembles can reach up to 0.85.

5.4 Hybrid Entanglement of Three Cold Atomic Ensembles and Three Flying Photons

During the realization of atom-atom entanglement, we acquire a final entanglement with a fidelity of 0.85 through optimizing the photon interference (entanglement swapping) and report the connection of two quantum network nodes. Going further, connection of three quantum network nodes can be also realized.

In order to achieve the goal, we firstly prepare three pairs of photon-atom entanglement, and then interfere the three photons to post-select the entanglement of three cold atomic ensembles and three flying photons. For the generated photon-atom entanglement, we measure the entanglement visibility under superpositional basis and overall retrieval efficiency in one of the read-out channels with adjusted storage time, the results are shown in Fig. 5.14. From Fig. 5.14a, the period of the oscillation curve is around 5.28 μs, which means once the time delay for Raman pulse changes 5.28 μs, the difference of accumulated phase between the atomic states $|\downarrow\rangle$ and $|\uparrow\rangle$ is 2π. Figure 5.14b shows the change of retrieval efficiency with increasing storage time, which gives a $1/e$ storage lifetime of 75(2) μs, mainly limited to the motion of spinwave [15]. The entanglement visibility exceeds the classical threshold $1/\sqrt{2}$ over 41 μs. When the source is applied to long distance entanglement, the storage lifetime needs to be further improved to sub-second level by trapping the atoms in a 3D optical lattice to confine atomic movements [16]. We calculate the entanglement fidelity as 0.926(4). After confirming the results conforms with the aforementioned result after optimization, we start the experiment connection of multiple cold atomic quantum memories.

Figure 5.15 shows the experimental setup, the three flying write-out photons come from three separate quantum network nodes, i.e., three cold atomic quantum memories. Two of the three photons come from two nodes marked as Node I and Node II located in the same lab, the third photon comes from Node III located in another lab and transmits through an 18 m fiber. The three write-out photons are sent into the interference device, detailed setup for optical path is shown in Fig. 5.16. The main change is to insert a PBS2 to couple the third photon for interference. The three pairs of photon-atom entanglement are written as:

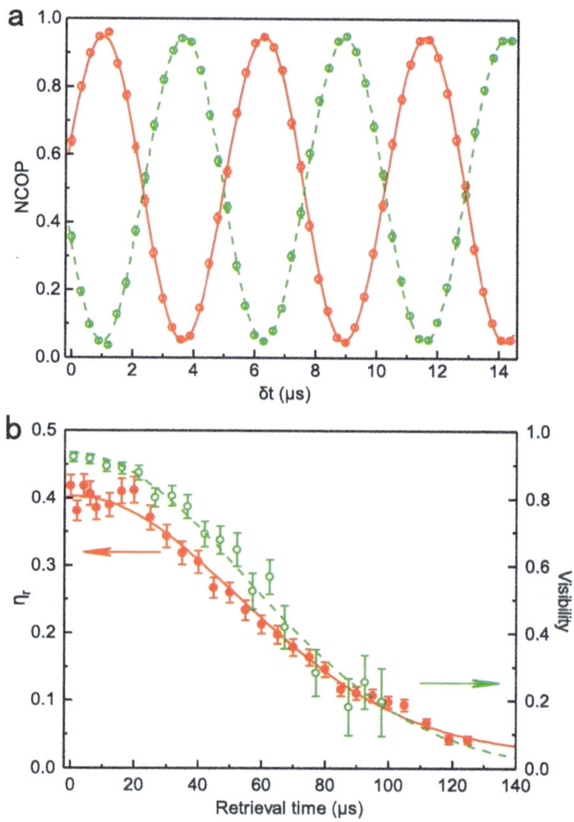

Fig. 5.14 Experimental results for single cold atomic quantum memory. **a.** Normalized coincidence probabilities (NCOP) as a function of Raman pulse delay δt relative to the write pulse, with the atomic state measured in the basis of $|\uparrow\rangle \pm |\downarrow\rangle$ and the write-out photon measured in the basis of $|\sigma^+\rangle \pm |\sigma^-\rangle$. Data points in red (solid line) correspond to parallel correlation, while data points in green (dashed line) correspond to cross correlation. Retrieval time is fixed to 15 μs. The oscillation has a period of 5.28 μs which is determined by the Zeeman splitting. **b.** Measured overall retrieval efficiency η_r and entanglement visibility in a superpositional basis as a function of retrieval time. An overall transmission efficiency of 0.65 and a detector efficiency of 0.70 are included in the overall retrieval efficiency

$$
\begin{aligned}
|\Psi\rangle_1 &= \frac{1}{\sqrt{2}}(|\sigma_1^+ \downarrow\rangle + |\sigma_1^- \uparrow\rangle), \\
|\Psi\rangle_2 &= \frac{1}{\sqrt{2}}(|\sigma_2^+ \downarrow\rangle + |\sigma_2^- \uparrow\rangle), \\
|\Psi\rangle_3 &= \frac{1}{\sqrt{2}}(|\sigma_3^+ \downarrow\rangle + |\sigma_3^- \uparrow\rangle),
\end{aligned}
\tag{5.47}
$$

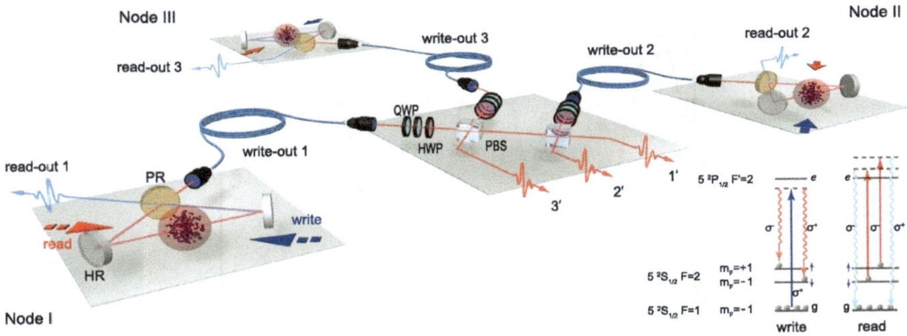

Fig. 5.15 Experimental setup for connection of three cold atomic quantum memories. Our experiment involves three quantum nodes. In each node, a setup of atomic-ensemble quantum memory with ring-cavity enhancement is employed. For each setup, Rubidium-87 atoms are initially prepared in a ground state $|g\rangle$, and a pair of atom-photon entanglement is created by making use of dual-rail spontaneous Raman scattering by applying a σ^+ polarized write beam coupling the transition $|g\rangle \leftrightarrow |e\rangle$. The write-out photons from three setups are sent to a middle station for entanglement connection via quantum interference. Atomic collective states are measured by applying a Raman pulse coupling the transition $|\uparrow\rangle \leftrightarrow |\downarrow\rangle$ for qubit rotation and applying a σ^- polarized read beam for optical retrieval and detection. To reduce noise due to leakage and Rayleigh scattered light from the control beams, we configure the write/read beams with detunings of different signs ($\mp$40 MHz). HWP: half-wave plate, QWP: quarter-wave plate, PBS: polarizing beam-splitter, PR: partially reflective mirror, HR: highly reflective mirror

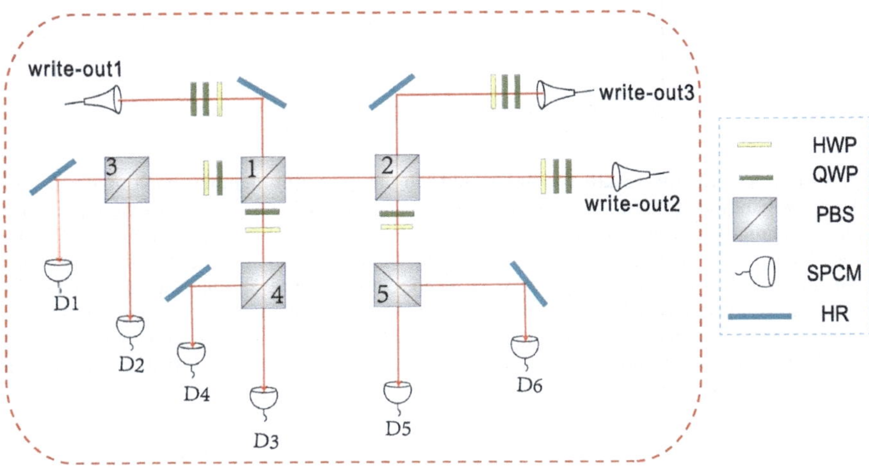

Fig. 5.16 Setup of optical path for interference of three write-out photons. PBS2 is placed on a linear translation stage for conveniently checking the two photon interference between "writeout1" and "writeout2"

where the subscript 1, 2, 3 represent the write-out photons from the three quantum memories. Before interference at the PBSs, we perform the polarization transformation as following: for the photons from Node I and Node II, $|\sigma^+\rangle \rightarrow |H\rangle$, $|H\rangle \rightarrow |H+V\rangle$, for the photons from Node III, $|\sigma^+\rangle \rightarrow |V\rangle$, $|H\rangle \rightarrow |H+V\rangle$. An extra requirement during the transformation is $|H\rangle \rightarrow |H+V\rangle$. The main purpose is to preserve the stability of relative phase between write-out photons with different polarizations when propagating through the optical fiber during the experiment and thus protect the entanglement from the influence of surroundings. Now the three pairs of photon-atom entanglement are:

$$
\begin{aligned}
|\Psi\rangle_1 &= \frac{1}{\sqrt{2}}(|H_1\downarrow\rangle + |V_1\uparrow\rangle),\\
|\Psi\rangle_2 &= \frac{1}{\sqrt{2}}(|H_2\downarrow\rangle + |V_2\uparrow\rangle),\\
|\Psi\rangle_3 &= \frac{1}{\sqrt{2}}(|V_3\downarrow\rangle + |H_3\uparrow\rangle).
\end{aligned}
\tag{5.48}
$$

The joint state for the tensor product of the three states is:

$$
\begin{aligned}
&|\Psi\rangle_1 \otimes |\Psi\rangle_2 \otimes |\Psi\rangle_3\\
&= \frac{1}{2\sqrt{2}}(|H_1H_2H_3\downarrow\downarrow\uparrow\rangle + |V_1V_2V_3\uparrow\uparrow\downarrow\rangle + |H_1H_2V_3\downarrow\downarrow\downarrow\rangle + |V_1V_2H_3\uparrow\uparrow\uparrow\rangle\\
&+ |H_1V_2H_3\downarrow\uparrow\uparrow\rangle + |V_1H_2V_3\uparrow\downarrow\downarrow\rangle + |V_1H_2H_3\uparrow\downarrow\uparrow\rangle + |H_1V_2V_3\downarrow\uparrow\downarrow\rangle).
\end{aligned}
\tag{5.49}
$$

The three flying write-out photons are sent into the PBS (1, 2) for interference, since the PBS transmits photons with H polarization and reflect photons with V polarization, if we only consider the case where there is one photon in each spatial mode after the two PBSs (in principle, the cases can be selected via quantum non-demolition measurement to check whether there exists one photon in the output ports of PBS and the quantum state will not be influenced), it means photons after PBSs are all H polarized or V polarized, thus the joint state in Eq. (5.49) will be projected into a sub-space consisting of $|HHH\downarrow\downarrow\uparrow\rangle$ and $|VVV\uparrow\uparrow\downarrow\rangle$. Through post-selecting the cases that there exists one photon in each output port of the PBS, the six fold entanglement of three cold atomic quantum memories and three flying write-out photons can be generated, the entangled state can be written as:

$$
|\mathrm{GHZ}_6\rangle = \frac{1}{\sqrt{2}}(|HHH\downarrow\downarrow\uparrow\rangle + |VVV\uparrow\uparrow\downarrow\rangle),
\tag{5.50}
$$

where the first three qubits represent the write-out photons (1′, 2′, 3′), the last three qubits represent the atomic spin state of the quantum memories. It can be also known that the probability to post-select the entangled state from the joint state in Eq. (5.49) is 25%. In order to verify the entanglement, we can measure the fidelity by decom-

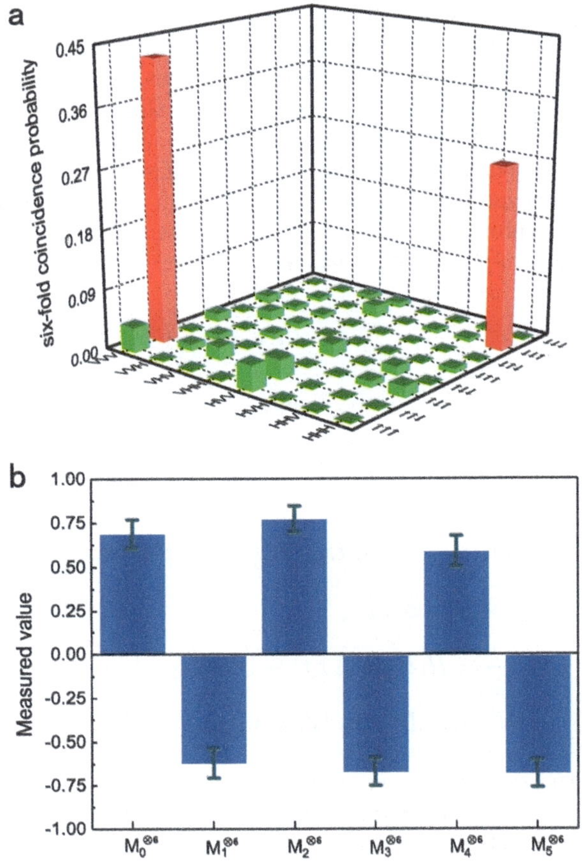

Fig. 5.17 Results for six-fold entanglement $|GHZ_6\rangle$. **a**, Normalized sixfold coincidence probabilities measured in the H/V basis for the flying write-out photons and in the $|\downarrow\rangle/|\uparrow\rangle$ basis for the atomic states. **b**, measured results for $M_n^{\otimes 6}$. Imbalance in the retrieval efficiency for different atomic states and different set-ups is calibrated

posing its density matrix into some observables that can be measured via some local measurements (See Appendix D):

$$\rho_6 = \frac{1}{2} M_{HV} + \frac{1}{12} \sum_{n=0}^{5} (-1)^n M_n^{\otimes 6}, \tag{5.51}$$

where $M_{HV} = |HHH \downarrow\downarrow\uparrow\rangle\langle HHH \downarrow\downarrow\uparrow| + |VVV \uparrow\uparrow\downarrow\rangle\langle VVV \uparrow\uparrow\downarrow|$ corresponds to the measurements under eigen basis, while $M_n = \cos(n\pi/6)\sigma_x + \sin(n\pi/6)\sigma_y$ corresponds to the measurement under superpositional basis. $\sigma_{x,y,z}$ are the Pauli matrices under the definition of eigen basis (σ_z) of $|H\rangle/|V\rangle$ for photons and $|\uparrow\rangle/|\downarrow\rangle$ for atoms. When measuring $M_0^{\otimes 6}$, the three QWPs after PBS 1, 2 are fixed to 45°, HWP is fixed to 22.5°, the time delay of Raman $\pi/2$ pulse is fixed to where the entanglement visibility is the maximum while measuring atom-atom entanglement, when measuring the $M_1^{\otimes 6}$, we add 7.5° to each HWP, and add 1/12 of the time period

Table 5.1 Measured value for M_{HV}, $M_n^{\otimes 6}$. It is worth to note that M_{HV} is the expected coincidences as a percentage of the total coincidences, $M_n^{\otimes 6}$ is the ratio of the difference between the expected coincidences and the undesired coincidences to the total coincidences

M_{HV}	$M_0^{\otimes 6}$	$M_1^{\otimes 6}$	$M_2^{\otimes 6}$	$M_3^{\otimes 6}$	$M_4^{\otimes 6}$	$M_5^{\otimes 6}$
0.702(41)	0.690(82)	−0.620(84)	0.773(74)	−0.670(79)	0.590(87)	−0.680(79)

(1/12×5.28 μs = 440 ns) in Node I and Node II, minus 440 ns (or add 4840 ns) in Node III to the time delay for applying Raman pulse. We repeat the operations in turn while measuring the remaining $M_2^{\otimes 6} - M_5^{\otimes 6}$. The measured results are shown in Fig. 5.17, Table 5.1, where the measured probability is 0.42 for $|VVV\downarrow\downarrow\uparrow\rangle$, 0.27 for $|HHH\uparrow\uparrow\downarrow\rangle$, 0.31 for other 62 state combinations.

Finally, the fidelity between measured state and expected state $|\mathrm{GHZ}_6\rangle$ can be calculated as:

$$\begin{aligned} F_6 &= \frac{1}{2}\langle M_{HV}\rangle + \frac{1}{12}\sum_{n=0}^{5}(-1)^n\langle M_n^{\otimes 6}\rangle \\ &= 0.686 \pm 0.026. \end{aligned} \tag{5.52}$$

which exceeds the classical threshold 0.5 about 7 standard deviations, sufficiently proving the six fold GHZ entanglement of three cold atomic ensembles and three write-out photons.

5.5 GHZ Entanglement of Three Cold Atomic Quantum Memories

Considering the state $|GHZ_6\rangle$, if we perform projective measurements on the three photons, the remaining state for three cold atomic quantum memories can be prepared to a GHZ state. This kind of entanglement of multiple quantum memories is the building block for performing quantum computing, realizing 2D quantum repeater and constructing quantum networks with cold atomic ensembles [1, 17]. The experimental setup for the entanglement is similar with that of six fold entanglement, where the main difference is to remove the QWPs after PBS 1, 2 in Fig. 5.16 and fix the HWP to 22.5°, thus the projective measurements on photons can be performed. During the experiment, we still select the events where there exists one photon in each output ports of PBS 1, 2. We deduce the non-normalized state transformation after all the PBSs as:

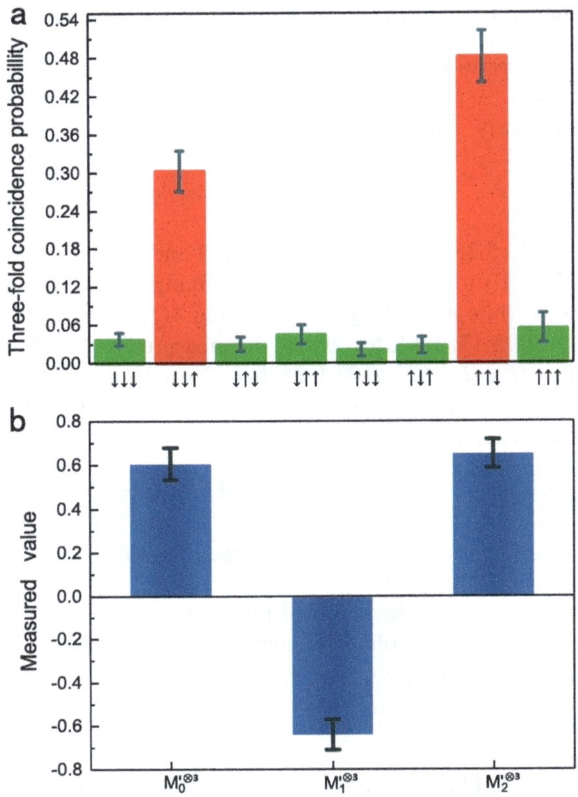

Fig. 5.18 Results for entanglement of three cold atomic ensembles. **a**, Normalized threefold coincidence probabilities measured in the $|\downarrow\rangle/|\uparrow\rangle$ basis for the atomic states. **b**, measured results for $M_n^{'\otimes 3}$. Imbalance in the retrieval efficiency for different atomic states and different set-ups is calibrated

Table 5.2 Measured value for M'_{HV}, $M_n^{'\otimes 3}$

	M'_{HV}	$M_0^{'\otimes 3}$	$M_1^{'\otimes 3}$	$M_2^{'\otimes 3}$
Measured value	0.785(33)	0.606(73)	−0.641(69)	0.651(66)

$$\begin{aligned}
|\Psi\rangle &\xrightarrow{\text{PBS 1\&2}} |HHH\downarrow\downarrow\uparrow\rangle + |VVV\uparrow\uparrow\downarrow\rangle \\
&\xrightarrow{\text{HWPs @ 22.5}^\circ} |DDD\downarrow\downarrow\uparrow\rangle + |AAA\uparrow\uparrow\downarrow\rangle \\
&= |HHH\rangle(|\downarrow\downarrow\uparrow\rangle + |\uparrow\uparrow\downarrow\rangle) + |HHV\rangle(|\downarrow\downarrow\uparrow\rangle - |\uparrow\uparrow\downarrow\rangle) + |HVH\rangle(|\downarrow\downarrow\uparrow\rangle - |\uparrow\uparrow\downarrow\rangle) \\
&+ |VHH\rangle(|\downarrow\downarrow\uparrow\rangle - |\uparrow\uparrow\downarrow\rangle) + |VVV\rangle(|\downarrow\downarrow\uparrow\rangle - |\uparrow\uparrow\downarrow\rangle) + |VVH\rangle(|\downarrow\downarrow\uparrow\rangle + |\uparrow\uparrow\downarrow\rangle) \\
&+ |VHV\rangle(|\downarrow\downarrow\uparrow\rangle + |\uparrow\uparrow\downarrow\rangle) + |HVV\rangle(|\downarrow\downarrow\uparrow\rangle + |\uparrow\uparrow\downarrow\rangle).
\end{aligned} \tag{5.53}$$

When coincident detections among SPCMs (1, 3, 5), (1, 4, 6), (2, 4, 5), (2, 3, 6) are measured, the state among the three cold atomic ensembles will collapse to:

$$|\text{GHZ}_3^+\rangle = \frac{1}{\sqrt{2}}(|\downarrow\downarrow\uparrow\rangle + |\uparrow\uparrow\downarrow\rangle). \tag{5.54}$$

When coincident detections among SPCMs (1, 3, 5), (2, 4, 6), (1, 4, 5), (2, 3, 5) are measured, the state among the three cold atomic ensembles will collapse to:

$$|\mathrm{GHZ}_3^-\rangle = \frac{1}{\sqrt{2}}(|\downarrow\downarrow\uparrow\rangle - |\uparrow\uparrow\downarrow\rangle). \tag{5.55}$$

As the evolution speed for atomic states $|\downarrow\rangle$ and $|\uparrow\rangle$ is different, we can always adjust the storage time to let $|\mathrm{GHZ}_3^-\rangle$ evolve to $|\mathrm{GHZ}_3^+\rangle$, which can improve the brightness of preparing entanglement of three cold atomic ensembles, which is the so-called feedfoward method.

To verify the quality of the entanglement, we also decompose the density matrix of this entangled state into some observables that can be obtained by simple local measurements,

$$\rho_3 = \frac{1}{2}M'_{HV} + \frac{1}{6}\sum_{n=0}^{2}(-1)^n M_n^{'\otimes 3}. \tag{5.56}$$

Similarly, $M'_{HV} = |\downarrow\downarrow\uparrow\rangle\langle\downarrow\downarrow\uparrow| + |\uparrow\uparrow\downarrow\rangle\langle\uparrow\uparrow\downarrow|$ represents the measurement under eigen basis, $M_n^{'\otimes 3} = \cos(n\pi/3)\sigma_x + \sin(n\pi/3)\sigma_y$ represents the measurement under superpositional basis. Figure 5.18, Table 5.2 show the results for measuring the atomic states after the three photons are performed projective measurements. As shown in Fig. 5.18a, the measured probability are 0.30 for $|\downarrow\downarrow\uparrow\rangle$, 0.48 for $|\uparrow\uparrow\downarrow\rangle$ and 0.22 for other six state combinations. When the atomic states are measured under superpositional basis, a single change of Raman pulse delay is adjusted from 1/12 period mentioned in the previous section to 1/6 period (880 ns), while other settings stay the same as aforementioned. Through the measured results, we calculate the fidelity of measured state compared to the ideal GHZ state as:

$$\begin{aligned} F_3 &= \frac{1}{2}\langle M'_{HV}\rangle + \frac{1}{6}\sum_{n=0}^{2}(-1)^n\langle M_n^{'\otimes 3}\rangle \\ &= 0.709 \pm 0.026. \end{aligned} \tag{5.57}$$

As a summary, we established the largest GHZ entangled state of three separate quantum memories for the first time in the world through the projection measurements on photons. The fidelity exceeds the classical threshold 0.5 by 8 standard deviations, sufficiently proving the three quantum memories are indeed entangled. The GHZ state of three separate quantum memories, acts as the basic building block for 2D quantum repeater.The importance of building a global quantum network is self-evident. Furthermore, since we measure photons by projection, we can predict whether the entanglement will be successfully generated through the measurement results. As it is now a probabilistic photon-atom entanglement source, three photons come from three quantum memories and two of the three photons come from the same quantum memories have the same probability, as a result, when the coincidence detection for three write-out photons is successful, there is about 1/7 probability that the three photons come from the three atomic ensemble respectively, so that the entanglement efficiency of the three cold atomic quantum memories is predicted successfully with about a probability of 1/7. This efficiency can be further increased to 100% by generating deterministic photon-atom entanglement in Rydberg atomic ensembles.

5.6 Efficiency of Entanglement Generation and Verification

In the previous section, we focus on the single-shot success probability of the prepared entangled state and the final fidelity. In order to complete this experiment, there are also requirements for the brightness that meets the photon counting rate in the experiment. If it is too low, it will cause the whole experiment difficult to carry out. Estimated with the results and time sequence in the previous experiment [7], it will take nearly 40 h to obtain a coincidence count in three fold GHZ entanglement experiment. This is obviously difficult to carry out experimentally and is of little significance to going further. To improve the counting rate, we not only optimize the efficiency of a single successful generation and verification of photon-atom entanglement, but also optimize the loading process of atoms, state preparation and other factors. The final repetition frequency that can be achieved for experimental trials is about 30 kHz. The detailed time sequences are as follows: The cycle time for the next time to load atoms is 21 ms, including 18 ms for the loading of atoms in the magneto-optical trap, optical molasses cooling and initial state preparation. In the next 3 ms, we carry out memory experiments and repeat the write process until the coincident detections of write-out photons, then we apply Raman pulse and read beam to verify the atomic state. The interval between two write beam with a pulse width of 50 ns is 4.7 μs, which can be repeated 622 times in total for each MOT loading. Before each write process, we use about depumper laser with 3 μs and σ-pump laser with 2 μs to again initialize the atomic state. The estimated photon counting rate in one hour is:

$$\frac{1\,\text{h}}{21\,\text{ms}} \times 622 \times 0.006^3 \times \frac{1}{4} \simeq 6, \tag{5.58}$$

which is consistent with the measured counting rate (5–6 per hour). The total measured time are 130 h for the six fold entanglement and 100 h for three fold entanglement.

For the six fold entanglement $|\text{GHZ}_6\rangle$, the entanglement generation efficiency is the probability of that photons $1'$, $2'$, $3'$ (Fig. 5.15) can all be collected into fibers. Considering the total efficiency to generate and collect a photon ($1'$, $2'$, $3'$) into fiber is 0.03, the post-selection efficiency for $|\text{GHZ}_6\rangle$ after the PBSs is 1/4, then we can calculate the generation efficiency of $|\text{GHZ}_6\rangle$ as:

$$\eta_{g6} = 0.03^3/4 = 6.75 \times 10^{-6}. \tag{5.59}$$

Thus the generated rate is:

$$R_{g6} = \eta_{g6} \cdot 30\,\text{kHz} \simeq 0.2\,\text{Hz}. \tag{5.60}$$

It means entangled state $|\text{GHZ}_6\rangle$ can be generated in each 5 s, comparable to the longest storage lifetime with cold atomic quantum memories. During the verification process, the detection efficiency for write-out photons is 0.68, the average measured efficiency for atomic states is 0.3, then the overall generation and verification efficiency for $|\text{GHZ}_6\rangle$ is:

$$\eta_6 = (0.03 \times 0.68 \times 0.3)^3/4 = 5.73 \times 10^{-8}. \tag{5.61}$$

Corresponding coincident counting rate is $5.73 \times 10^{-8} \times 30\,\text{kHz} = 0.0017\,\text{Hz} \simeq 6$ counts/hour, consistent with the experimental results.

Similarly, for the three fold GHZ state $|\text{GHZ}_3\rangle$, the generation efficiency is probability of the successful projection measurements on the three photons, the post-selection efficiency for the state is 1/8 and can be improved to 1/4 with the feed-fowrad method by controlling the storage time. When the three write-out photon are successfully detected, the generation efficiency of GHZ state for atoms can be calculated as $(0.03 \times 0.68)^3/4 = 2.12 \times 10^{-6}$. Combined with the verification efficiency for the atomic states, the overall efficiency for generation and verification of $|\text{GHZ}_3\rangle$ is the same with $|\text{GHZ}_6\rangle$, i.e., 5.73×10^{-8}. The final coincident counting rate is also around 6 coincidences per hour.

5.7 Summary

In this chapter, based on previous experiments, combined with the improvement in the performance of high-efficiency retrievable quantum memory in the experiment, the process of generating photon-atom entanglement is carefully studied, and the factors affecting its fidelity are analyzed to obtain high fidelity entanglement source. Based on these achievements, we have thoroughly studied the various parameters that need to be optimized to realize the entanglement of two cold atomic ensembles. In particular, we have made considerable progress in analyzing the indistinguishability of different photons, and thus lay the foundations for the connection of multiple quantum network nodes. Finally, after optimizing the photon-atom entanglement source and realizing high-fidelity photon interference, we realize for the first time entanglement of three separate cold atomic ensembles, six fold entanglement of three photons and three separate cold atomic ensembles. The entanglement provides a very favorable condition for the further research of quantum network, quantum computing, quantum precision measurement and so on.

References

1. Wallnöfer J, Zwerger M, Muschik C, Sangouard N, Dür W (2016) Two-dimensional quantum repeaters. Phys Rev A 94(5):052307
2. Matsukevich DN, Kuzmich A (2004) Quantum state transfer between matter and light. Science 306(5696):663–666
3. Chou C-W, De Riedmatten H, Felinto D, Polyakov SV, Van Enk SJ, Kimble HJ (2005) Measurement-induced entanglement for excitation stored in remote atomic ensembles. Nature 438(7069):828
4. Duan L-M, Lukin MD, Cirac JI, Zoller P (2001) Long-distance quantum communication with atomic ensembles and linear optics. Nature 414(6862):413
5. Sangouard N, Simon C, De Riedmatten H, Gisin N (2011) Quantum repeaters based on atomic ensembles and linear optics. Rev Mod Phys 83(1):33
6. Chen S, Chen Y-A, Zhao B, Yuan Z-S, Schmiedmayer J, Pan J-W (2007) Demonstration of a stable atom-photon entanglement source for quantum repeaters. Phys Rev Lett 99(18):180505
7. Yang S-J, Wang X-J, Li J, Rui J, Bao X-H, Pan J-W (2015) Highly retrievable spin-wave-photon entanglement source. Phys Rev Lett 114(21):210501
8. Clauser JF, Horne MA, Shimony A, Holt RA (1969) Proposed experiment to test local hidden-variable theories. Phys Rev Lett 23(15):880
9. Saffman M, Walker TG, Mølmer K (2010) Quantum information with rydberg atoms. Rev Modern Phys 82(3):2313
10. Li J, Zhou M-T, Jing B, Wang X-J, Yang S-J, Jiang X, Mølmer K, Bao X-H, Pan J-W (2016) Hong-ou-mandel interference between two deterministic collective excitations in an atomic ensemble. Phys Rev Lett 117(18):180501
11. Yuan Z-S, Chen Y-A, Zhao B, Chen S, Schmiedmayer J, Pan J-W (2008) Experimental demonstration of a bdcz quantum repeater node. Nature 454(7208):1098
12. Li L, Dudin YO, Kuzmich A (2013) Entanglement between light and an optical atomic excitation. Nature 498(7455):466
13. Pan J-W, Chen Z-B, Chao-Yang L, Weinfurter H, Zeilinger A, Żukowski M (2012) Multiphoton entanglement and interferometry. Rev Mod Phys 84(2):777
14. Zhao T-M, Zhang H, Yang J, Sang Z-R, Jiang X, Bao X-H, Pan J-W (2014) Entangling different-color photons via time-resolved measurement and active feed forward. Phys Rev Lett 112(10):103602
15. Zhao B, Chen Y-A, Bao X-H, Strassel T, Chuu C-S, Jin X-M, Schmiedmayer J, Yuan Z-S, Chen S, Pan J-W (2009) A millisecond quantum memory for scalable quantum networks. Nat Phys 5(2):95
16. Yang S-J, Wang X-J, Bao X-H, Pan J-W (2016) An efficient quantum light-matter interface with sub-second lifetime. Nat Photon 10(6):381
17. Barrett SD, Rohde PP, Stace TM (2010) Scalable quantum computing with atomic ensembles. New J Phys 12(9):093032

Chapter 6
Interference of Three Frequency Distinguished Photons

When connecting different physical platforms, it often encounters the interference of non-identical photons. In this chapter, I will mainly introduce how to use the time-resolved measurement method to achieve the interference of non-identical photons with completely distinguishable frequencies, and obtain the information of the interference network through the interference results. Our work will help lay the foundation for the establishment of quantum networks among different physical systems. It also provides a method for using non-identical photon interference to carry out Bose sampling experiments and illustrate quantum advantages.

6.1 Background

In the previous chapter, we report the entanglement of three quantum network nodes, the basic building block for a global quantum network, it lays the foundation to extend more quantum network nodes. During the connection of multiple quantum network nodes, one of the most important step is high-fidelity photon-photon interference, however, the interfered photons often come from different nodes (maybe different physical systems) and will not be identical, the interference results based on identical photons thus will be influenced and finally lead to a decreasing fidelity of node-node entanglement. In consideration of this issue, it is meaningful to study interference of distinguishable photons. Here, I will focus on the interference of photons that are distinguishable in frequency modes, the work will lay the foundation for future compatibility with photon interference among different physical systems. At the same time, it also lays the foundation for the realization of the Bose sampling experiment with distinguishable input states [1, 2] and the quantum computation based on non-identical photon interference. It also provides a research direction for quantum supremacy with the use of interaction of distinguishable single photons.

B. Jing, *Quantum Network with Multiple Cold Atomic Ensembles*, Springer Theses,
https://doi.org/10.1007/978-981-19-0328-1_6

6.2 Single Photon Source

In order to perform interference of multiple photons, the first step is to prepare multiple single photons. Similar with generation of correlated photon pairs with spontaneous parametric down-conversion (SPDC) process, where detection of one photon is used as the heralded signal for the existence of the other photon, in our experiment, we use the write-out photons as the heralded signal for the single read-out photons. The difference is that the photon pairs are generated at the same time in SPDC process while the read-out photons can be released on demand after detection of the write-out photons. Since the retrieval efficiency has been improved to around 0.9, it is high enough for us to make use of the process for efficiently generating single photons. The experimental setup is shown in Fig. 4.12, ring-cavity enhanced quantum memories with cold atomic ensembles are used for efficient generation of single read-out photons. The used atomic energy levels and corresponding write/read process are shown in Fig. 6.1. Since the whole process does not involve the entanglement of photonic polarizations and atomic states, we can use the two write-out channels to generate the same read-out photons to increase the brightness of the single-photon source. The operation is described as follows: all atoms are initially pumped into $|F = 1, m_F = -1\rangle$, after applying the write beam with polarization of σ^+, when the write-out photon is σ^+ polarized, the corresponding atomic state is $|F = 2, m_F = -1\rangle$, then the read beam with polarization of σ^- is applied directly to release a single read-out photon with polarization of σ^-. When the write-out photon is σ^- polarized, the corresponding atomic state is $|F = 2, m_F = +1\rangle$, here we perform a Raman π pulse to transfer the atom to state $|F = 2, m_F = -1\rangle$, then still use the σ^- polarized read beam to generate the read-out photons. The single read-out photons obtained in the two processes have the same optical properties, but the atoms scattered to other energy levels are used and can double the brightness of the single photon source.

Since the waveshape of the single photon will also affect the result of the interference, especially the difference in frequency will also affect the distribution of the

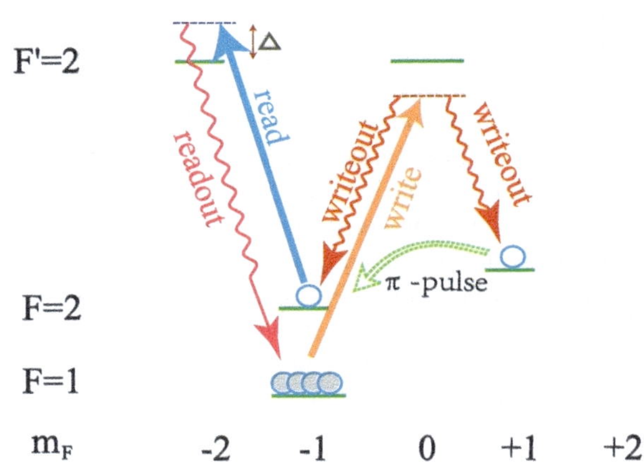

Fig. 6.1 Schematic for generation of single read-out photons

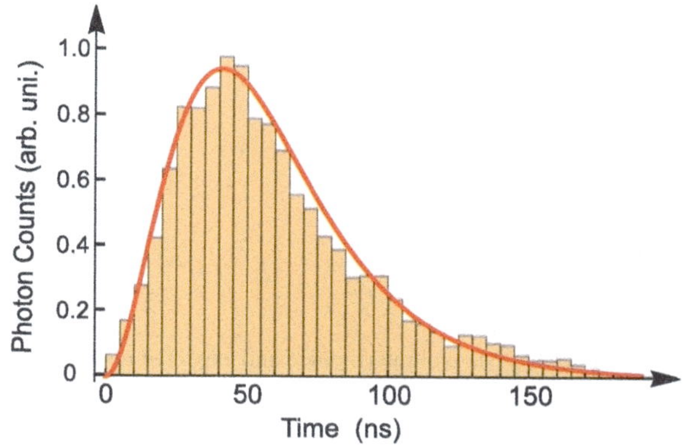

Fig. 6.2 Temporal distribution of single read-out photons. The fitting function is $[Erf(t/\tau)Exp(-t/\tau)]^2$, when using the equation to analyze the results of multi-photon interference, it will be impossible to obtain an analytical solution. Therefore, in the early analysis, the Gaussian function is usually used to approximately simulate the waveform

detection time, here we first measure the time distribution of the read-out photons that will be used in the experiment, as shown in Fig. 6.2.

6.3 HOM Interference Between Two Indistinguishable Single Photons

In 1987, Chung Ki Hong et al. firstly present the Hong-Ou-Mandel interference [3], where photons will go into the same output port if two identical photons are sent into a beam splitter simultaneously from the two input ports for perfect interference, as shown in Fig. 6.3. Photons are sent into input ports 1 and 2, then the probability of detection of coincident event between output ports 3 and 4 would be zero. The process can be explained as following: the state for input photons can be written as:

$$|1_1\rangle|2_1\rangle = a_1^\dagger a_2^\dagger|vac\rangle, \tag{6.1}$$

while the BS performs a transformation on the photonic states as:

$$\begin{aligned} a_1^\dagger &\rightarrow \frac{1}{\sqrt{2}}(a_4^\dagger + ia_3^\dagger) \\ a_2^\dagger &\rightarrow \frac{1}{\sqrt{2}}(a_3^\dagger + ia_4^\dagger). \end{aligned} \tag{6.2}$$

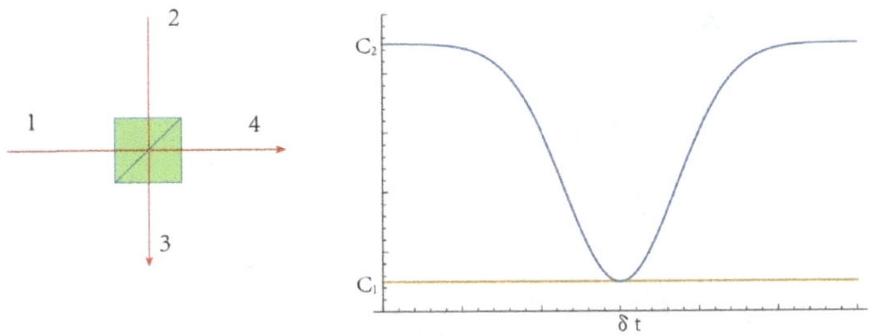

Fig. 6.3 Schematic for Hong-Ou-Mandel interference and expected results

The photonic states after the BS thus become:

$$\begin{aligned} a_1^\dagger a_2^\dagger |vac\rangle &= \frac{1}{2}(a_4^\dagger + i a_3^\dagger)(a_3^\dagger + i a_4^\dagger) \\ &= \frac{i}{2}(a_3^{\dagger 2} + a_4^{\dagger 2})|vac\rangle. \end{aligned} \tag{6.3}$$

It can be seen that there will be no two photon coincidences between the two output ports after the interference of two identical photons at the BS. If the two input photons are not sent at the same time, for instance, when they are completely separated in time, the two photons pass through the BS without any interference. At this time, the coincidence counts can be observed between the two output ports. By adjusting the time delay between the two input photons, the two-photon coincidence counts will satisfy the varying trend shown in Fig. 6.3. When the photons arrive at the interference at the same time, the coincidence count is the least, and the valley that appears here is called the HOM dip. In the experiments, the visibility of HOM interference is usually defined to assess the indistinguishability of two photons. The coincidence count measured when two photons are completely distinguishable is C_2, and the coincidence count measured when two photons completely interfere is C_1, the visibility is then defined as[1]:

$$V = \frac{2C_2}{C_1}. \tag{6.4}$$

If there exists some difference between the photons, for example, the two photons differs in polarization, one has a polarization of H and the other one has a polarization of $cos\theta H + sin\theta V$, after the BS, the photonic states becomes:

$$\begin{aligned} a_1^\dagger &= \frac{1}{\sqrt{2}}(i a_3^\dagger H_3 + a_4^\dagger H_4) \\ a_2^\dagger &= \frac{1}{\sqrt{2}} a_3^\dagger (cos\theta H_3 + sin\theta V_3) + \frac{1}{\sqrt{2}} i a_4^\dagger (cos\theta H_4 + sin\theta V_4), \end{aligned} \tag{6.5}$$

[1] A percent definition in other references may be $(C_2 - C_1)/C_2$.

then we have:

$$a_1^{\dagger}a_2^{\dagger}|vac\rangle = \frac{1}{2}[ia_3^{\dagger 2}H_3^2cos\theta + ia_4^{\dagger 2}H_4^2cos\theta + ia_3^{\dagger 2}H_3V_3sin\theta \\ + a_3^{\dagger}a_4^{\dagger}H_4V_3sin\theta - a_3^{\dagger}a_4^{\dagger}H_3V_4sin\theta + ia_4^{\dagger 2}H_4V_4sin\theta]|vac\rangle. \tag{6.6}$$

The final detected probability for the coincidences is:

$$p_{34} = \frac{1}{2}sin^2\theta = \frac{C_1}{2C_2}. \tag{6.7}$$

In our experiment, we use cold atomic quantum memory to generate single photons, and use the single read-out photons heralded by the detected write-out photons to perform two-photon interference. To measure the similarity of photons except for the degree of freedom of frequency, we first generate read-out photons with the same frequency and perform the HOM interference. The experimental results are shown in Fig. 6.4.

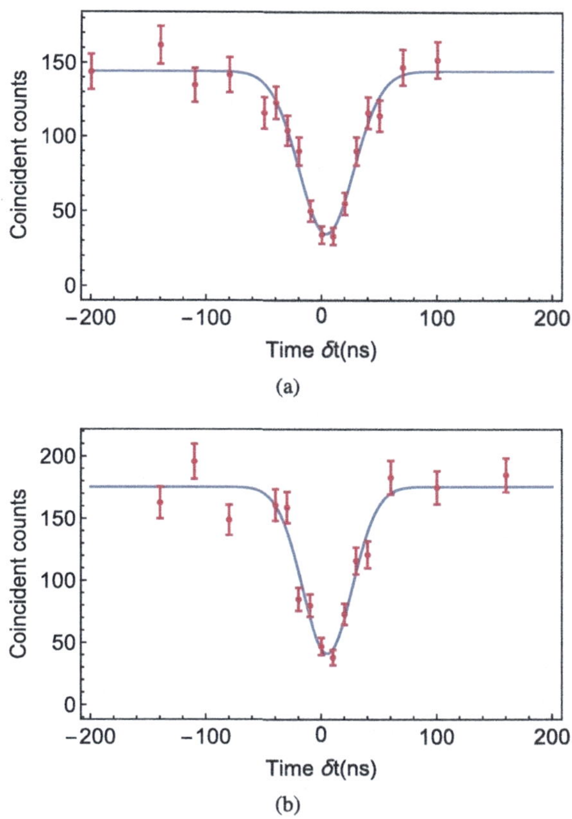

Fig. 6.4 Results for HOM interference of read-out photons. **a**, result for HOM interference of two read-out photons with excitation probability of 0.01. **b**, result for HOM interference of two read-out photons with excitation probability of 0.02. The time delay for photons arriving at the BS is adjusted through the storage time

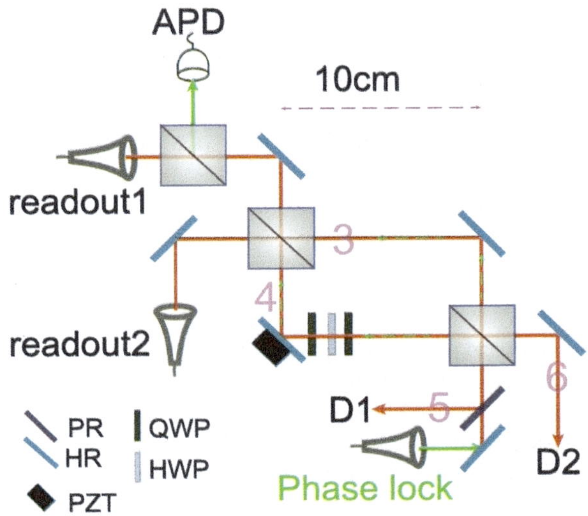

Fig. 6.5 Experimental setup for two photon interference

The interference visibility is around 9:1. As the excitation probability of a single write channel is usually about 0.01, even after doubling, it can be seen that the HOM interference visibility has hardly changed, indicating that the single-photon purity of the read-out photon is not the main factor affecting the interference in that situation. After photon interference with single BS, the next step is to start the two-photon network experiment. The experimental setup is shown in Fig. 6.5. Conditioned on the detection of write-out photons, the read-out photons named as readout1 and readout2 from the two quantum memories are adjusted to V polarized and sent into a MZ interferometer. In order to make the interferometer have the best stability, here we try to build an interferometer with a small size, as can be seen from the figure, each arm of the interferometer has a length of about 16 cm. The phase of the interferometer is locked by a phase-locked beam (polarization of H) that is sent into the interferometer in the opposite direction. The phase-locked beam is coupled into the optical path through a partial reflective mirror (PR) with a reflectivity of 0.95 and the signal after the interferometer is detected by an APD and feedback to the PZT is given by a circuit board to stabilize the phase of the two arms of the interferometer. Since the polarization of the experimental light is orthogonal to that of the phase-locked beam, in order to allow the phase of the interferometer experienced by the experimental light to be adjusted arbitrarily, one half-wave plate and two quarter-wave plates are applied to adjust the relative phase difference between the two lights. While the phase-locked beam can always lock the interferometer to zero phase, phase differences can be obtained when the read-out photon passes through the two arms of the interferometer. Here we adjust the phase difference of the read-out photons to $\pi/2$, so that after the read-out photons 1 and 2 pass through the interferometer, there will be no coincidence counts detected. Then we adjust the time difference between the read-out photons and measure the HOM interference, the result is shown in Fig. 6.6.

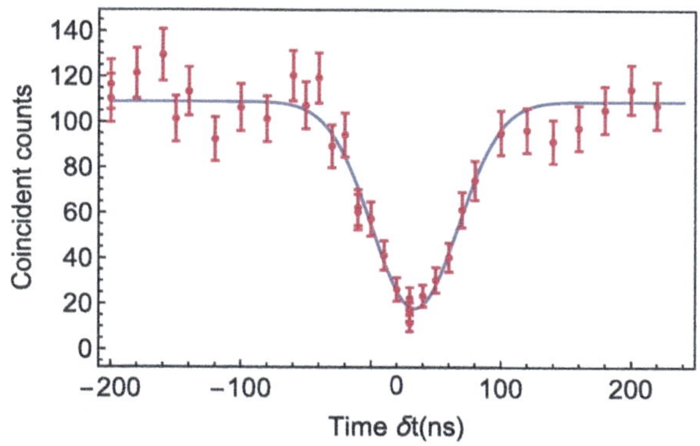

Fig. 6.6 Results for photon interference with MZ interferometer

6.4 Interference of Two Frequency Distinguished Photons

When two indistinguishable photons pass through a BS and interfere, they tend to go to the same output port after the BS, while for distinguishable photons, they will go to the two output ports of the BS randomly, however, even the frequency of two photons differs in real situation, the photons still interfere when passing through the BS. We can use a more sensitive detector to erase the different frequency information carried by the photons, as a result, after the frequency-distinguished photons passing through the BS, we can still detect their coincident events at the two output ports. The coincident countings can still give the interfered information [4].

Here, we take two photons with Gaussian beam profile as an example. The two input photons have an angular frequency difference of $\Delta = 2\pi \times 20\,\text{MHz}$, their wavepacket functions can be described as:

$$\begin{aligned} f_1(t) &= e^{-(t/\delta\tau)^2 - i\omega t} \\ f_2(t) &= e^{-(t/\delta\tau)^2 - i(\omega+\Delta)t} \end{aligned} . \tag{6.8}$$

Here $\delta\tau$ is set to 50 ns, after the BS (Fig. 6.3), the creation operator after output ports 3 and 4 satisfies:

$$\begin{aligned} a_3^\dagger(t) &= \frac{1}{\sqrt{2}}[ia_1^\dagger f_1(t) + a_2^\dagger f_2(t)] \\ a_4^\dagger(t) &= \frac{1}{\sqrt{2}}[a_1^\dagger f_1(t) + ia_2^\dagger f_2(t)] \end{aligned} . \tag{6.9}$$

Considering the detection events that one photon at output port 3 is detected at time t_1 and the other photon at output port 4 is detected at time t_2, the joint detection probability then can be calculated as:

$$p_{34}(t_1, t_2) = \langle a_3(t_1)a_4(t_2)|a_1^\dagger a_2^\dagger\rangle^2. \tag{6.10}$$

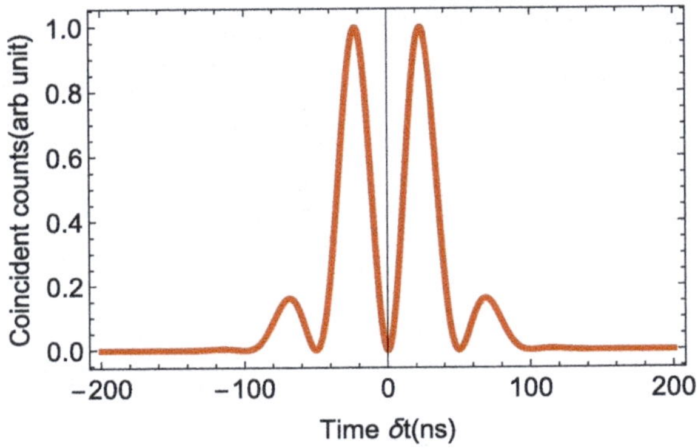

Fig. 6.7 Theoretical calculation results for the photon coincidences versus the relative detection time

If the relative detection time $\delta t = t_2 - t_1$ is the main factors we consider, then we can rewrite the probability as:

$$P_{34}(\delta t) = \int dt_1 \langle a_3(t_1) a_4(t_1 + \delta t) | a_1^\dagger a_2^\dagger \rangle^2. \tag{6.11}$$

The theoretical calculation results are shown in Fig. 6.7. As seen from the figure, the coincidence counts distribution of the two output ports of the BS presents an obvious interference pattern with the detection time difference determined by the two photons. When the detection time for both photons is the same, the disappearance of the coincidence can still be observed. At this time, the photons still tend to go to the same output ports of the BS, i.e., the perfect HOM interference result. Even the coincidences are randomly distributed if the entire detection window is averaged, the detection of photons by the fast detector can erase the distinguishable information of the two frequencies, so that the interference of photons can still be observed. The period of the interference pattern corresponds exactly to the reciprocal of the frequency difference between the two photons.

Furthermore, we sent the two frequency distinguished photon into a MZ interferometer. In order to improve the brightness and scalability of the experiment, we do not generate read-out photons every time we finish the write process, because this is no different from a low-probability single-photon source, and the scalability will be poor when the number of photons increases. Given that our quantum memory can store the atomic states for a certain period of time and then read them out, we can independently repeat the write operation for each quantum memory until the write-out photon is detected and stop. When write-out photons from all quantum memories have been successfully detected, it indicates that there is already a spin wave stored in each atomic ensemble. At this time, the read beam is applied at the same time to convert the atomic states into single read-out photons. Thanks to the high retrieval efficiency, the probability of obtaining multiple single photons at the same time will be much higher than that with parametric down conversion process. Meanwhile, it can be predicted whether there are read-out photons by detection of write-out photons. When the number of photons increases, the probability of prepar-

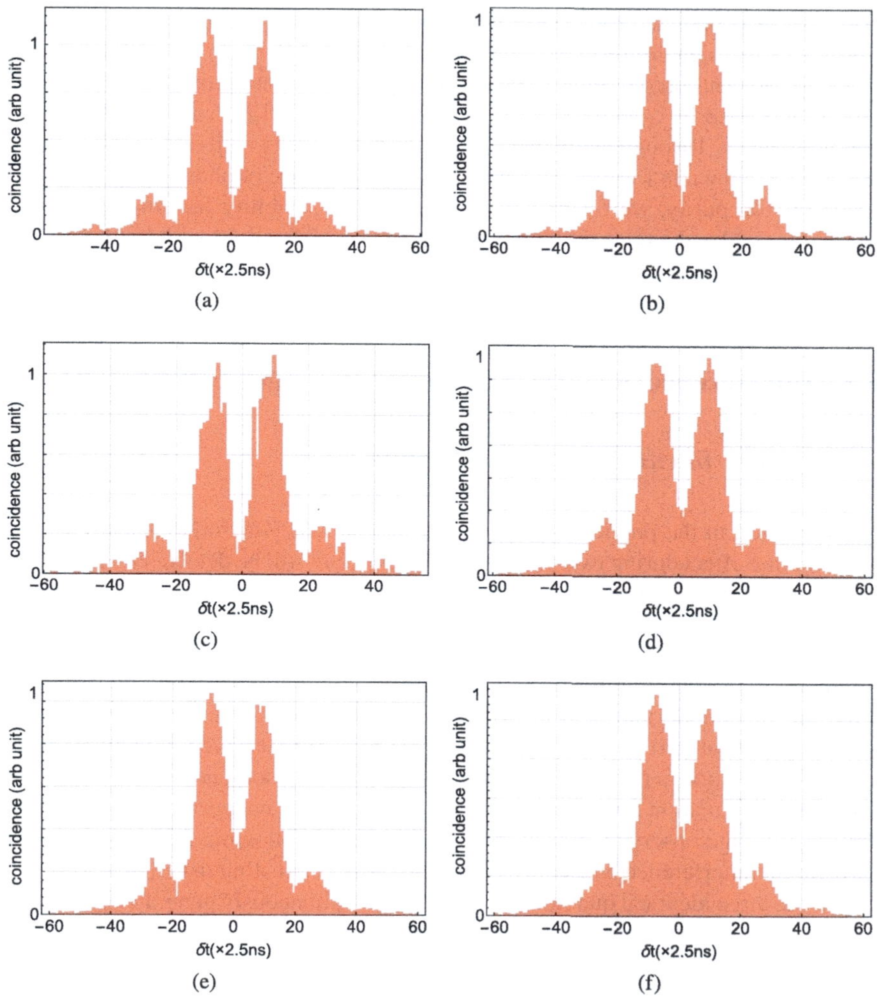

Fig. 6.8 Results for two photon interference under different excitation probability. The excitation probability for each write-out channel is 0.01, 0.018, 0.02, 0.035, 0.042, 0.06 shown in (**a**)–(**f**), respectively

ing multiple photons at the same time also changes from scaling as the exponential decrease in parametric down conversion to the current polynomial level. If the storage time is long enough, we can simply repeat the write process until all ensembles have successfully prepared spin waves, and then read them out to acquire single read-out photons at the same time. However, considering the finite storage lifetime around 64 μs, we set a maximum number $m = 7$ of attempts for the write process, if detection of write-out photons from all quantum memories succeeds within m times, the read process is operated for the following experiment, if it has reached m times and there still exists quantum memory that no write-out photon is detected, then we start a new experiment preparation. By doing this, the n-photon preparation probability can be improved by a factor of $[1-(1-p_e)^m]^n/(mp_e^n)$ [5], here p is the excitation probability. Through analyzing the interference results with different excitation probability (shown in Fig. 6.8) and considering the balance of photon brightness and single photon purity, we finally set the excitation probability for single write-out channel to 0.02 and a total excitation probability of 0.04 is used with both write-out channels.

6.5 Interference of Three Frequency Distinguished Photons

6.5.1 Experimental Setup and Results

As mentioned in the previous section, even for frequency-distinguishable photons, the bunching effect during two-photon interference can still be detected by fast single photon detectors. This detection method is the so-called time-resolved measurement, and there have been groups reported the perfect entanglement swapping operation with interference of frequency distinguished photons through time-resolved measurement [6]. Recently, Vincenzo Tamma et al. found that when using photons with different frequencies and polarizations for Bose sampling experiments, if the photons at the output ports of the linear optical network are measured with polarization and time resolution, a richer multiphoton interference phenomenon can be observed, and some information about the symmetry of the optical network itself can be also derived from the observed interference pattern [7, 8]. Following Ref. [7], we firstly perform the interference experiment of three frequency distinguished photons.

We use three identical quantum memories to simultaneously generate three single read-out photons. The linear optical network is similar to that shown in Fig. 6.5. The difference is that we will add an additional 2:1 BS. The experimental setup is shown Fig. 6.9 [5], the frequencies of the three single photons are achieved by adjusting the frequency of read beam and cavity of each memory respectively. The corresponding frequency for each read beam is set to blue detuned from the excited energy levels with detunings of $2\pi \times 72.4$ MHz (s_1), $2\pi \times 33.0$ MHz (s_2) and $2\pi \times 52.4$ MHz (s_3), respectively. The generated single read-out photons are all Fourier transform-limited and the coherence time is mainly determined by the laser linewidth, i.e., about 1 MHz, far exceeding the resolved time of single photon detector and detection time window, which can enable the observation of three photon interference with time resolved measurement. If the photon coherence time is smaller than the detection time window, three photons interference results will no longer be able to be obtained through time-resolved measurements.

The phase φ in the path 6 is adjusted through two QWPs and one HWP, which can result in phase adjustment on the single photons by locking the interferometer via

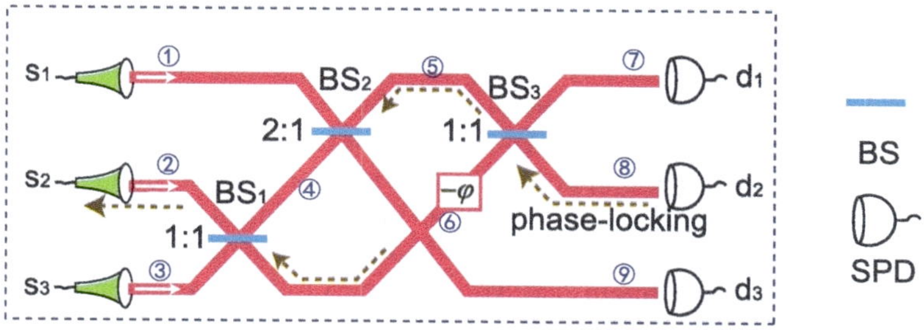

Fig. 6.9 Linear optical network for three photon interference

locking beam transmitted on the contrary direction and with polarization orthogonal to the single photons. Through analyzing the interference, the interference pattern will present different symmetries by varying the phase φ. We firstly deduce the transformation relationships of the three-photon interference network as:

$$\begin{aligned}
\xi_1(t) &= f_1(t)e^{-i\omega_1 t}a_1^\dagger, \\
\xi_2(t) &= f_2(t)e^{-i\omega_2 t}a_2^\dagger, \\
\xi_3(t) &= f_3(t)e^{-i\omega_3 t}a_3^\dagger \\
\xi_4(t) &= \frac{1}{\sqrt{2}}[\xi_3(t) + i\xi_2(t)], \\
\xi_5(t) &= \frac{1}{\sqrt{3}}[\xi_4(t) + 2i\xi_1(t)], \\
\xi_6(t) &= \frac{1}{\sqrt{2}}[\xi_3(t) + i\xi_3(t)]e^{-i\varphi}, \\
\xi_7(t) &= \frac{1}{\sqrt{2}}[\xi_6(t) + i\xi_5(t)], \\
\xi_8(t) &= \frac{1}{\sqrt{2}}[\xi_5(t) + i\xi_6(t)], \\
\xi_9(t) &= \frac{1}{\sqrt{3}}[\xi_1(t) + i\xi_4(t)].
\end{aligned} \tag{6.12}$$

Here, $f_{1,2,3}(t)$ represent the temporal distribution of photonic wavepacket, during the experiment, the detection time, corresponding to $\xi_7(t_1)$, $\xi_8(t_2)$, $\xi_9(t_3)$ in the aforementioned equations, of photons in the three output ports are recorded. The transformation matrix of the 3×3 optical network is described as:

$$U(\varphi) = \frac{1}{\sqrt{3}}\begin{bmatrix} -1 & (\sqrt{3}e^{-i\varphi} - 1)/2 & i(\sqrt{3}e^{-i\varphi} + 1)/2 \\ i & i(\sqrt{3}e^{-i\varphi} + 1)/2 & (1 - \sqrt{3}e^{-i\varphi})/2 \\ 1 & -1 & i \end{bmatrix}.$$

When $\varphi = 0$, the permanent of the matrix is zero, i.e., $Per[U(0)] = 0$. This means that the three identical photons have a bunching effect after passing through the optical network, and no three-photon coincidences with each output port having one photon will be observed. When $\varphi = \pi/2$, each photon has an equal probability going into one of the three output ports, i.e., with an equal probability of 1/3, the optical network is then called a *tritter* [9]. In our experiment, the frequency difference of input photons is larger than the photonic linewidth, the frequency of three photons are completely distinguished. Through time resolved measurement with fast detector, the information of frequency difference can be erased and let us observe the phenomena of no coincidences with $\varphi = 0$ and the maximum coincident probability with a tritter.

In order to be able to compare the experimental results with the theoretical analysis, the matrix for optical network needs to be reconstructed. Usually, to reconstruct an unknown optical network matrix, it is necessary to utilize the output result with both single-photon incidence and two-photon incidence, respectively. However, as the single photons come from the read process of the DLCZ quantum memory, its frequency is resonant with the cavity and also depends on the frequency of read beam. It is not convenient enough to adjust the frequency. Considering in our experiment, the main uncertain factors are the BS and the phase φ in the optical path, we can reconstruct the network matrix via the likelihood methods with only one photon sending into the network [10]. In detail, we parameterize the network parameters into the phase φ and reflectivity and transmittance of the BSs. The photons passing through the network have the following transformations:

$$\begin{aligned}
A_5 &= A_4\sqrt{T_2} + iA_1\sqrt{R_2};\\
A_9 &= A_1\sqrt{T_2} + iA_4\sqrt{R_2};\\
A_4 &= A_3\sqrt{T_1} + iA_2\sqrt{R_1};\\
A_7 &= A_6\sqrt{T_3} + iA_5\sqrt{R_3};\\
A_8 &= A_5\sqrt{T_3} + iA_6\sqrt{R_3};\\
A_6 &= e^{-i\varphi}A_2\sqrt{T_1} + iA_3\sqrt{R_1},
\end{aligned} \tag{6.13}$$

where R and T is the reflectivity and transmittance of the BSs, respectively. The network matrix are parameterized as:

$$\mathcal{M}_k = \begin{pmatrix} -\sqrt{p_{11}R_2R_3} & \sqrt{p_{12}}\left(e^{-i\varphi}\sqrt{T_1T_3} - \sqrt{R_1T_2R_3}\right) & \sqrt{p_{13}}\left(ie^{-i\varphi}\sqrt{R_1T_3} + i\sqrt{T_1T_2R_3}\right) \\ i\sqrt{p_{21}R_2T_3} & \sqrt{p_{22}}\left(i\sqrt{R_1T_2T_3} + ie^{-i\varphi}\sqrt{T_1R_3}\right) & \sqrt{p_{23}}\left(\sqrt{T_1T_2T_3} - e^{-i\varphi}\sqrt{R_1R_3}\right) \\ \sqrt{p_{31}T_2} & -\sqrt{p_{32}R_1R_2} & i\sqrt{p_{33}T_1R_2} \end{pmatrix}. \tag{6.14}$$

Here, p_{mn} represents the overall detected efficiency for photons in output port m with input photons in port n. We send the single photons from the three input ports in turn, and measure the photon events in three output ports to acquire the distribution probability. The measured results are as follows:

$$E_0 = \begin{pmatrix} 0.08424 & 0.01573 & 0.15029 \\ 0.08446 & 0.16960 & 0.01083 \\ 0.10274 & 0.09577 & 0.08718 \end{pmatrix}; \tag{6.15}$$

$$E_1 = \begin{pmatrix} 0.08479 & 0.09043 & 0.08683 \\ 0.08380 & 0.09523 & 0.08140 \\ 0.10084 & 0.09669 & 0.08864 \end{pmatrix}; \tag{6.16}$$

$$E_2 = \begin{pmatrix} 0.08489 & 0.17028 & 0.01399 \\ 0.08581 & 0.01386 & 0.15502 \\ 0.10309 & 0.09543 & 0.08796 \end{pmatrix}; \tag{6.17}$$

$$E_3 = \begin{pmatrix} 0.08542 & 0.09439 & 0.07792 \\ 0.08337 & 0.08973 & 0.08774 \\ 0.10264 & 0.09395 & 0.08604 \end{pmatrix}. \tag{6.18}$$

$E_{k,mn}$ represents overall retrieval efficiency for detection of single read-out photons conditioned on the write-out photons when the phase is set to $\varphi = \frac{\pi}{2}k(k = 0, 1, 2, 3)$ and single photons sent into the network from the nth input port are measured in the mth output port. When analyzing the matrix, what affects is the relative efficiency among different input-output configurations. To make the parameterized matrix mostly approximate to the measured results, we use a likelihood function defined as:

$$f = \sum_{k=0}^{3} \sum_{m=1}^{3} \sum_{n=1}^{3} \left| |\mathcal{M}_{k,mn}|^2 - E_{k,mn} \right|^2. \tag{6.19}$$

By solving the function to find its minimum value, we can fit the parameters in $\mathcal{M}_k$ and acquire the matrix as following:

$$\mathcal{M}_0 = \begin{pmatrix} -0.5617 & 0.2435 + 0.0128i & -0.0119 + 0.7498i \\ 0.5626i & -0.0124 + 0.7950i & -0.2056 - 0.0119i \\ 0.6177 & -0.5974 & 0.5723i \end{pmatrix}, \tag{6.20}$$

$$\mathcal{M}_1 = \begin{pmatrix} -0.5606 & -0.2526 - 0.5197i & 0.4855 + 0.2854i \\ 0.5615i & 0.5059 + 0.3111i & 0.2552 + 0.4827i \\ 0.6166 & -0.5963 & 0.5712i \end{pmatrix}, \tag{6.21}$$

$$\mathcal{M}_2 = \begin{pmatrix} -0.5576 & -0.7867 - 0.0809i & 0.0756 - 0.2164i \\ 0.5584i & 0.0788 - 0.2118i & 0.7512 + 0.0752i \\ 0.6132 & -0.5931 & 0.5681i \end{pmatrix}, \tag{6.22}$$

$$\mathcal{M}_3 = \begin{pmatrix} -0.5607 & -0.2833 + 0.5204i & -0.4861 + 0.2569i \\ 0.5616i & -0.5065 + 0.2813i & 0.2837 - 0.4833i \\ 0.6167 & -0.5964 & 0.5714i \end{pmatrix}. \quad (6.23)$$

Here the matrix has been normalized by dividing a coefficient around $\sqrt{0.27}$ related to p_{mn}, 0.27 is also the product of single photon generation efficiency (0.45) and overall transmission and detection efficiency (0.6).

In our experiment, after the three write-out photons from three cold atomic quantum memories are all detected, the read beams are applied to convert the stored spinwaves into three single read-out photons. The three photons are then sent to the interference network, and photons leaving the network are measured with single photon detectors and a time-to-digital converter (U1051A, Agilent, Inc.) is used to record the photon detection time t_1, t_2, t_3. Here we focus on the coincident events with relative detection time, and count all events with the same time difference $x \equiv t_1 - t_3$, $y \equiv t_2 - t_3$. For each data at (x_i, y_j), we again use the data of its surrounding points (x, y) to perform a weighted average according to the Gaussian weight distribution, the weight function is defined as:

$$w(x, y) = e^{-[(x-x_i)^2+(y-y_i)^2]/r_0^2}. \quad (6.24)$$

r_0 is set to 3 ns in consideration of the resolution of the processed figure and the fluctuation of the counts. Figure 6.10 shows the three photon coincident events with different detection time difference x, y.

As clearly seen from the figure, when $\varphi = 0, \pi$, if the detection time difference is zero, there exists no coincidences, in consistent with the permanent of zero for the network matrix. Furthermore, through Fourier analysis, there exists a oscillation similar with a quantum beat in the x axis, the corresponding period is 51.4(11) ns, which is caused by the interference of photons 1 and 3 whose frequency difference is 20 MHz. The period corresponding to the oscillation in y axis is 24.9(3) ns, also consistent with the frequency difference between photon 1 and 2, i.e., 39.4 MHz.

When $\varphi = \pi/2$, the three photon interference network is called *tritter*, which can be simply described with a matrix as $U_{ds} = e^{-i2\pi ds/3}/\sqrt{3}$ with $s = 1, 2, 3$ represent the input photon modes and $d = 1, 2, 3$ represent the output photon modes. The interference result is different from that with $\varphi = 0$, in the center of $x = y = 0$, the photon coincident events are no longer zero, on the contrary, it corresponds to the situation where the information on frequency difference has been erased, the output results should be that three photons respectively output to three output ports or all enter one output port, coincidences of three photons in the three output ports thus can be observed. Regardless of the coincidences in the center, the period of the x direction and the y direction can be analyzed to be 49.7(7) ns and 50.3(13) ns.

To further estimate the measured result, we also deduce a theoretical result and make a comparison. The theoretical result is shown in Fig. 6.10e–h. In general, the experimental results are basically consistent with the theoretical analysis results. In order to quantitatively describe the degree of similarity between the experimental

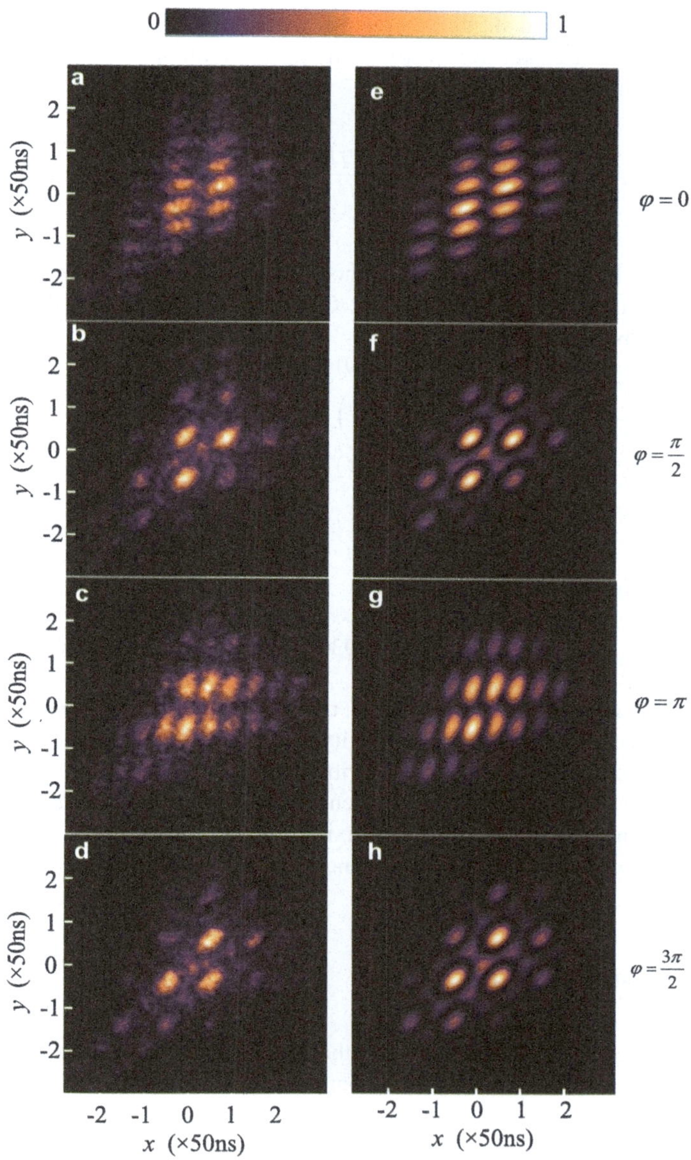

Fig. 6.10 Time-resolved coincidences of three photon interference with different phase. The left side shows the experimental results while the right side shows the theoretical results with the reconstructed matrix. The horizontal axis is the time difference $t_1 - t_3$ between the photon detection time t_1 with detector d_1 and the time t_3 with the detector t_3, the vertical axis is the time difference $t_2 - t_3$ between the photon detection time t_2 with detector d_2 and the time t_3 with the detector t_3

results and the theoretical simulation results, we make use of the probability distribution in the two cases and follow the statistically commonly used Bhattacharyya distance to define the fidelity as:

$$F = \frac{\sum_{i,j}^{N,M} f_e\left(x_i, y_j\right)^{1/2} \cdot f_t\left(x_i, y_j\right)^{1/2}}{\left[\sum_{i,j}^{N,M} f_e\left(x_i, y_j\right)\right]^{1/2} \cdot \left[\sum_{i,j}^{N,M} f_t\left(x_i, y_j\right)\right]^{1/2}}, \tag{6.25}$$

where $f_e(x, y)$ is the measured coincidence distribution (Fig. 6.10a–d), $f_t(x, y)$ is the expected coincidence distribution (Fig. 6.10e–h). The calculated fidelity with the four setting phases is:

$$\begin{aligned} F(\varphi = 0) &= 0.946, \\ F\left(\varphi = \frac{\pi}{2}\right) &= 0.930, \\ F(\varphi = \pi) &= 0.953, \\ F(\varphi = \frac{3\pi}{2}) &= 0.925. \end{aligned} \tag{6.26}$$

An average fidelity is thus acquired:

$$\bar{F} = 0.939(13). \tag{6.27}$$

Going further, when $\varphi = \pi/2$ or $3\pi/2$, the interference network is a tritter, there should be unique symmetry in the probability distribution of the detected coincidence counts [8], here, we directly plot the distribution with three detection time t_1, t_2, t_3, as shown in Fig. 6.11. It can be clearly seen from the figure that the distribution has obvious rotational symmetry around the axis (1, 1, 1). The measured symmetry can be also used to infer that the interference network should be a tritter.

6.5.2 Fidelity Analysis

In the experiment, the fidelity between the experimental interference results and theoretical analysis is 0.939. One of the main fidelity limitations is the single photon quality of the read-out photons. Since the read-out photons are not strictly single photons, the multiple excitation events in the storage process will cause the read-out photons to contain two-photons or even more photons, which will affect their single-photon qualities, and thus affect the final result of the measurement. To obtain the ratio of multi-photon events, we firstly measure the single photon quality function α with varying storage time in each quantum memory, as shown in Fig. 6.12. As the repeating write process is set to a maximum value of 7, the longest storage time we needed is 35 μs, within this limit, the single photon quality function nearly stays constant, the average value for those measured α within 35 μs is shown in Table 6.1.

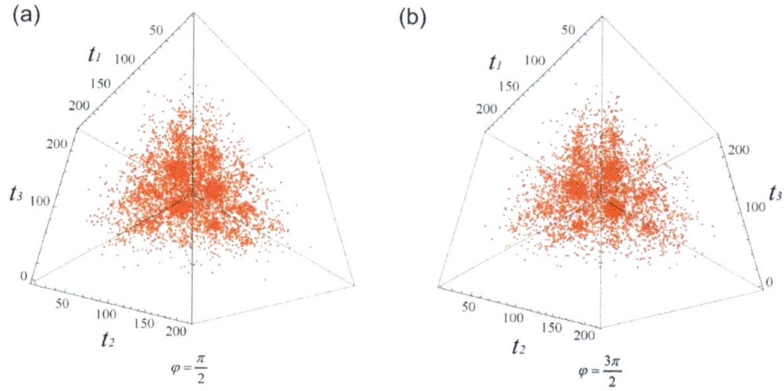

Fig. 6.11 3D distribution of the detected coincidence counts with detection time for each photon when the interference network is a tritter. **a**, $\varphi = \pi/2$. **b**, $\varphi = 3\pi/2$. The viewing angle is along the axis (1, 1, 1)

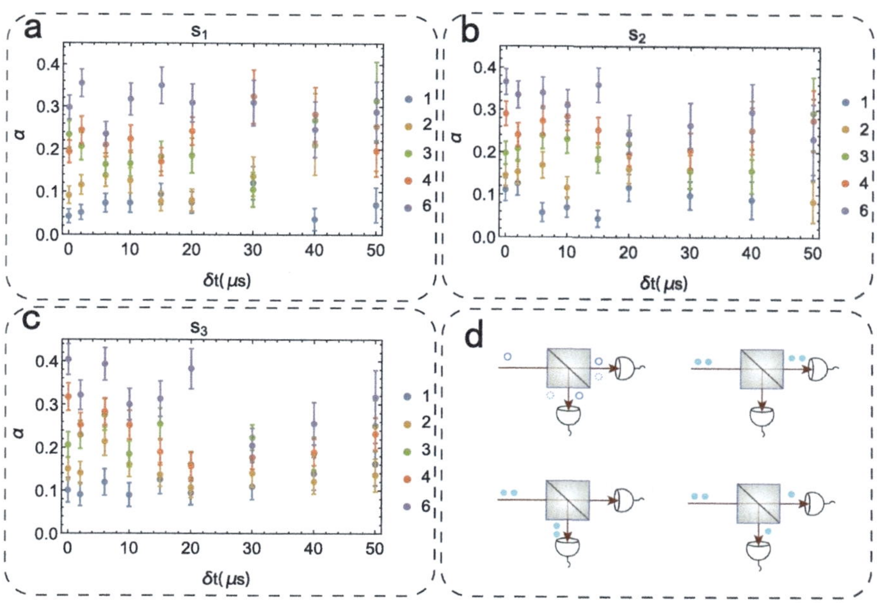

Fig. 6.12 Measured results of α. **a–c**, the measured α for three single read-out photons from the three cold atomic quantum memories varying with the storage time. The legends represent $p_e \times 100$ with p_e the excitation probability. **d**, the setup to measure the α with one photon or two photons input

Table 6.1 The average single photon quality function α over experimental time less than 35 μs with different excitation probability

p_e	0.01	0.02	0.03	0.04	0.06
$\alpha(s_1)$	0.072(7)	0.126(9)	0.201(11)	0.233(11)	0.322(12)
$\alpha(s_2)$	0.094(9)	0.165(10)	0.222(11)	0.279(11)	0.335(13)
$\alpha(s_3)$	0.110(9)	0.142(8)	0.220(10)	0.251(10)	0.361(13)

Here we mainly consider the influence of two-photon events. Assuming the probability is p_1 for read-out photons containing only one photon and p_2 for read-out photons containing two photons. p_1 and p_2 satisfy $p_1 + p_2 = 1$. As shown in Fig. 6.12d, if there only one photon sent in, no coincidences between output ports of BS will be observed. The probability to detect a photon in either output port is $\frac{1}{2}p_1$. Is there are two photons sent in, the probability is $\frac{1}{4}p_2$ for both photons going into the same output port, while the probability is $\frac{1}{2}p_2$ for the two photons going into different output ports. Then the probability for either output port having a photon is $\frac{1}{2}p_1 + \frac{3}{4}p_2$, the measured α should be:

$$\alpha = \frac{\frac{1}{2}p_2}{\left(\frac{1}{2}p_1 + \frac{3}{4}p_2\right)^2}. \tag{6.28}$$

The excitation probability in the experiment is set to 0.04 and the corresponding α is measured as 0.25, thus we can estimate the probability of read-out photons containing only one photon is 0.86 and the probability of read-out photons containing two photons is 0.14. Because of the existence of two-photon inputs, the measured interference results after the optical network differs from the ideal case where only one photon exists in each input port, as a result, the measured fidelity will be influenced. Here, we analyze the influence on fidelity mainly considering ideal situation A and two imperfect situations B & C, as shown in Fig. 6.13.

A: each input and output port exist one photon.

B: two photons at input port si, one photon at sj and no photons at the rest input port $(i, j = 1, 2, 3, i \neq j)$.

C: two photons at input port si, one photon at each of the other two input ports $sj, sk (i, j, k = 1, 2, 3, i \neq j \neq k, j > k)$.

Following Ref. [4], the electric field operators at input ports are described as:

$$E_{si}^{+}(t) = \zeta_{si}(t, \omega_{si}) a_{si}, \tag{6.29}$$

where $\zeta_{si}(t, \omega_{si})$ is the spatio-temporal mode function describing photon wavepacket. Given the transfer matrix $\mathcal{U}$, field operators at three output ports d1, d2, d3 are,

$$\{E_{d1}^{+}(t), E_{d2}^{+}(t), E_{d3}^{+}(t)\}^T = \mathcal{U}\{E_{s1}^{+}(t), E_{s2}^{+}(t), E_{s3}^{+}(t)\}^T. \tag{6.30}$$

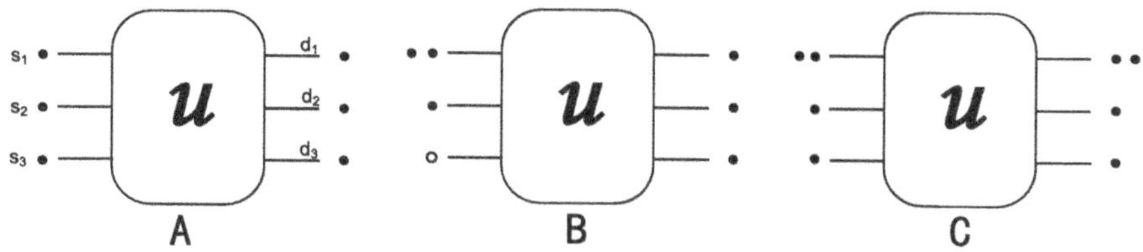

Fig. 6.13 Different combinations of injecting photons. Solid circle represents one photon and empty circle means no photon here

In ideal situation A, the input state is:

$$|\Psi_{in1}\rangle = a_{s1}^{\dagger} a_{s2}^{\dagger} a_{s3}^{\dagger} |vac\rangle. \tag{6.31}$$

To obtain the probability for photon detections in output ports d1, d2, d3 at times $t_1 = t_3 + \delta t_1$, $t_2 = t_3 + \delta t_2$, t_3, we have to apply the field operators $E_{d1}^{\pm}(t_1)$, $E_{d2}^{\pm}(t_2)$, $E_{d3}^{\pm}(t_3)$ on the input state, the joint probability is calculated as,

$$P_{co1}(\delta t_1, \delta t_2) = \int_{-\infty}^{+\infty} dt_3 \langle \Psi_{in1}| E_{d3}^{-}(t_3) E_{d2}^{-}(t_2) E_{d1}^{-}(t_1) E_{d1}^{+}(t_1) E_{d2}^{+}(t_2) E_{d3}^{+}(t_3) |\Psi_{in1}\rangle. \tag{6.32}$$

Conditioned on detection of write-out photons, average probability of existence of read-out photons at input port is $P_{ro} = 0.45$, where the probability is $P_1 = 0.86$ for only one photon and $P_2 = 0.14$ for existence of two photons deduced from an average $\alpha = 0.25$. The probability to detect a photon after the network is $P_d = 0.6$ where overall transmission efficiency and detector efficiency are included. The final probability distribution to detect three photons with different time in this situation would be:

$$P_{co1}'(\delta t_1, \delta t_2) = P_{ro}^3 P_1^3 P_d^3 P_{co1}(\delta t_1, \delta t_2), \tag{6.33}$$

which can be acquired through numerical methods and corresponds to the coincidences distribution in ideal cases.

In situation B, the input state is:

$$|\Psi_{in2}\rangle = a_{si}^{\dagger} a_{si}^{\dagger} a_{sj}^{\dagger} |vac\rangle. \tag{6.34}$$

Similarly, to obtain the probability for photon detections in output ports d1, d2, d3 at times $t_1 = t_3 + \delta t_1$, $t_2 = t_3 + \delta t_2$, t_3, we also apply the field operators $E_{d1}^{\pm}(t_1)$, $E_{d2}^{\pm}(t_2)$, $E_{d3}^{\pm}(t_3)$, the joint probability would be,

$$\begin{aligned} P_{co2}(\delta t_1, \delta t_2) = \sum_{i,j} \int_{-\infty}^{+\infty} & dt_3 \langle \Psi_{in2}| E_{d3}^{-}(t_3) E_{d2}^{-}(t_2) E_{d1}^{-}(t_1) \\ & E_{d1}^{+}(t_1) E_{d2}^{+}(t_2) E_{d3}^{+}(t_3) |\Psi_{in2}\rangle. \end{aligned} \tag{6.35}$$

Taking the probability of photon generation, transmission and detection into account, the final probability distribution to detect three photons in this situation is:

$$P'_{co2}(\delta t_1, \delta t_2) = P_{ro}^2(1 - P_{ro})P_1 P_2 P_d^3 P_{co2}(\delta t_1, \delta t_2). \quad (6.36)$$

In situation C, the input state is

$$|\Psi_{in3}\rangle = a_{si}^{\dagger} a_{si}^{\dagger} a_{sj}^{\dagger} a_{sk}^{\dagger} |vac\rangle, \quad (6.37)$$

which have 3 different combinations with different i, j, k. For the output ports, two photons at di, one photon at $dj, dk(i, j, k = 1, 2, 3, i \neq j \neq k, j > k)$, which also have 3 combinations. Here, for instance, two photons at $d1$ with photon detected time $t_1 = t_3 + \delta t_1$, $t_4 = t_3 + \delta t_4$, one photon at $d2$ with detected time $t_2 = t_3 + \delta t_2$, one photon at $d3$ with detected time t_3. Similar with Eq. 6.35, the coincident probability would be

$$\begin{aligned} P_{co3}(\delta t_1, \delta t_2, \delta t_4) = \sum_{i,j} \int_{-\infty}^{+\infty} & dt_3 \langle \Psi_{in3} | E_{d1}^{-}(t_4) E_{d3}^{-}(t_3) E_{d2}^{-}(t_2) E_{d1}^{-}(t_1) \\ & E_{d1}^{+}(t_1) E_{d2}^{+}(t_2) E_{d3}^{+}(t_3) E_{d1}^{+}(t_4) | \Psi_{in3} \rangle. \end{aligned} \quad (6.38)$$

As our recording system would only record photons detected first at detected channel with two photons that arrive at nearly the same time, there exists two cases: I. the first photon is detected and time $\min(\delta t_1, \delta t_4)$ is recorded with probability P_d. II. the first photon isn't detected while the latter one is detected and time $\max(\delta t_1, \delta t_4)$ is recorded with probability $(1 - P_d)P_d$. The final probability distribution at two cases are:

$$\begin{aligned} P'_I(\delta t_2, \min(\delta t_1, \delta t_4)) &= P_{ro}^3 P_1^2 P_2 P_d^2 P_d P_{co3}(\delta t_1, \delta t_2, \delta t_4) \\ P'_{II}(\delta t_2, \max(\delta t_1, \delta t_4)) &= P_{ro}^3 P_1^2 P_2 P_d^2 P_d(1 - P_d) P_{co3}(\delta t_1, \delta t_2, \delta t_4). \end{aligned} \quad (6.39)$$

The probability distributions obtained by the other various input and output distributions can also be obtained by the same method, and finally the distributions under the above three situations can be obtained by numerical methods, as shown in Fig. 6.14. It can be seen that situation B affects the most.

Based on the joint probability distribution of three situations analyzed above, we estimate the fidelity from an ideal case 1–0.976. Considering much higher-order events, the fidelity would become worse, so we briefly owe the main limitations of fidelity to the single photon quality function α. Other imperfections such as difference of pulse shape, phase instability of the linear optic network, and laser frequency instability will also contribute to the infidelity.

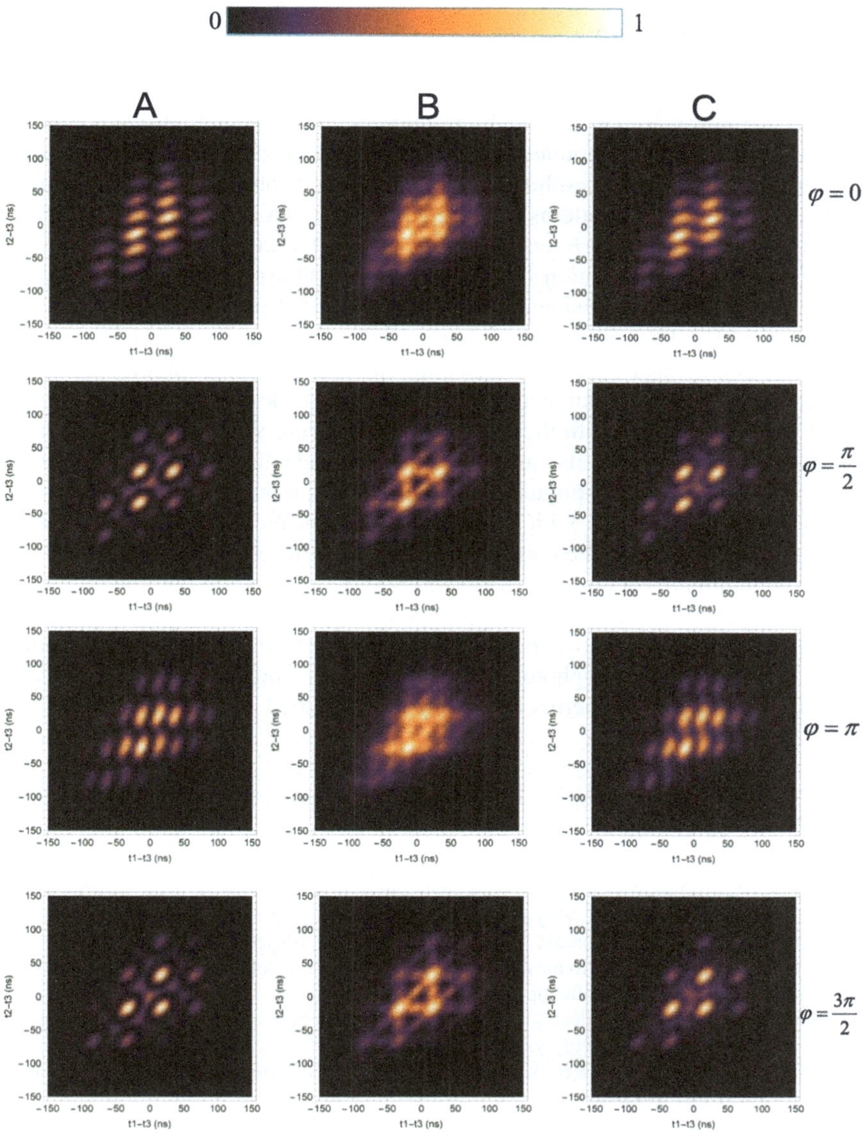

Fig. 6.14 Probability distribution diagram of three-photon coincidences with different situations analyzed under experimental conditions

6.6 Summary

In this chapter, I take the connection of different physical systems and Bose sampling as the experimental background and introduce the experimental advances from the interference of identical photons to the multi-photon interference with distinguishable frequencies. In the three-photon interference results obtained in the experiment, a wealth of interference patterns is obtained, and the symmetry of the interference patterns caused by the interference network with different symmetries is analyzed, which can also give information about the symmetry of the network vice versa. Moreover, in our experiment, the single photon source comes from quantum memory, and its brightness has been improved through the feedback mode. When the number of photons is extended to more photons, the multi-photon preparation time can be reduced from the exponential increase under parametric down-conversion process to the polynomial increase with the use of quantum memory. Through the utilization of quantum memories, it is also more advantageous than direct frequency shifting in producing photons with distinguishable frequencies but the same considering other parameters. Based on this, with the help of quantum memories, our work lays a foundation for improving the compatibility among different physical systems, connecting quantum network nodes with different systems, and also lays the foundation for realizing interference of more different photons through more complex networks. Meanwhile, it also provides a direction for the use of time-resolved measurement technology to study multi-photon interference and to prove quantum advantages through Bose sampling experiments among distinguished photons.

References

1. Aaronson S, Arkhipov A (2011) The computational complexity of linear optics. In: Proceedings of the forty-third annual ACM symposium on Theory of computing. ACM, pp 333–342
2. Gard BT, Motes KR, Olson JP, Rohde PP, Dowling JP (2015) An introduction to boson-sampling. In: From atomic to mesoscale: the role of quantum coherence in systems of various complexities. World Scientific, pp 167–192
3. Hong C-K, Zhe-Yu O, Mandel L (1987) Measurement of subpicosecond time intervals between two photons by interference. Phys Rev Lett 59(18):2044
4. Legero T, Wilk T, Kuhn A, Rempe G (2003) Time-resolved two-photon quantum interference. Appl Phys B 77(8):797–802
5. Wang X-J, Jing B, Sun P-F, Yang C-W, Yong Yu, Tamma V, Bao X-H, Pan J-W (2018) Experimental time-resolved interference with multiple photons of different colors. Phys Rev Lett 121(8):080501
6. Zhao T-M, Zhang H, Yang J, Sang Z-R, Jiang X, Bao X-H, Pan J-W (2014) Entangling different-color photons via time-resolved measurement and active feed forward. Phys Rev Lett 112(10):103602
7. Tamma V, Laibacher S (2015) Multiboson correlation interferometry with arbitrary single-photon pure states. Phys Rev Lett 114(24):243601

8. Laibacher S, Tamma V (2017) Symmetries and entanglement features of inner-mode resolved correlations of interfering nonidentical photons. arXiv:1706.05578
9. Weihs G, Reck M, Weinfurter H, Zeilinger A (1996) Two-photon interference in optical fiber multiports. Phys Rev A 54(1):893
10. Tillmann M, Dakić B, Heilmann R, Nolte S, Szameit A, Walther P (2013) Experimental boson sampling. Nat Photonics 7(7):540

Chapter 7
Entanglement of Two Cold Atomic Ensembles via 50 km Fibers

Since photons at telecom band have the lowest fiber transmission loss, the use of light at this band when transmitting optical signals is beneficial to enhance the entanglement distance between quantum memories, but nowadays many quantum memories are not easy to generate quantum correlations with telecom band photons directly. For our experiments, the solution is to use quantum frequency conversion to convert the near-infrared photons to telecom band photons for transmission. Therefore, in this chapter I will focus on the combination of quantum frequency conversion techniques with photon-atom entanglement to increase the photon transmission distance and realize the entanglement between two cold atomic ensembles via 50 km fibers.

7.1 Background

The distance between two adjacent nodes in a quantum network is an important parameter for realistic applications. If the distance is too short, then it will not only require higher quality of the entangled source when connecting more distant nodes, but also increase the consumption of resources, so the distance between adjacent quantum nodes should be enhanced as much as possible. In Chap. 5, we implemented the world's first GHZ entangled state containing three separate quantum network nodes, which is a fundamental building block for realizing two-dimensional quantum repeater and building quantum networks, but right now the entangled state is still only generated within the laboratory distance, and there is still a long way to go to connect quantum network nodes that are farther apart. One of the main reasons for this lies in the fact that the experimentally generated light compatible with the cold atomic quantum memory is in the near-infrared wavelengths, and for light in the near-infrared wavelengths, even in fiber optic transmission, the loss is as high as near 3 dB/km (Fig. 1.2), making it difficult to transmit over long distances. For photons at the telecom band, the fiber loss can be as low as 0.2 dB/km, so

B. Jing, *Quantum Network with Multiple Cold Atomic Ensembles*, Springer Theses,
https://doi.org/10.1007/978-981-19-0328-1_7

converting NIR photons to telecom band photons by quantum frequency conversion can reduce the loss in long-distance transmission, thus making long-distance quantum communication feasible. For the system we are studying, there are two main ways to perform quantum frequency conversion, one is based on the fact that the energy level structure of the Rb atom itself contains energy level transitions in the telecom band, and the NIR to telecom band conversion can be achieved by using the four-wave mixing effect within the atomic system [1, 2]. The other is through the differential frequency effect within the crystal or waveguide [3, 4]. In our experiments, we use a PPLN (periodically-poled lithium niobate) waveguide to convert photons at 795 nm to 1342 nm for long-distance transmission, and finally achieve the entanglement of two cold-atom quantum memories at a fiber distance of 50 km from each other. This is also the longest transmission distance for entanglement between quantum memories ever achieved.

7.2 Experimental Principle and Setup

Due to the limitations of the current experimental conditions, entanglement between different spin states inside the memory would result in low brightness of the entangled states, so we first carry out entanglement between the Fock states of the atomic spin states, in line with the basic principle of the Ref. [5]. The key point of the experiment is to increase photon transmission distance between quantum memories to 50 km by combining the efficiently retrievable quantum memory and quantum frequency conversion techniques.

Figure 7.1 shows the experimental setup, the atomic energy levels used for quantum memories of the two cold atomic ensembles (MOT_1 and MOT_2) are $|1, -1\rangle$ and $|2, -1\rangle$. The generated two pairs of Fock-type photon-atom entanglement can be written as:

$$|\Psi\rangle_{ap} = |0\rangle_a|0\rangle_p + \sqrt{p}|1\rangle_a|1\rangle_p + o(p), \tag{7.1}$$

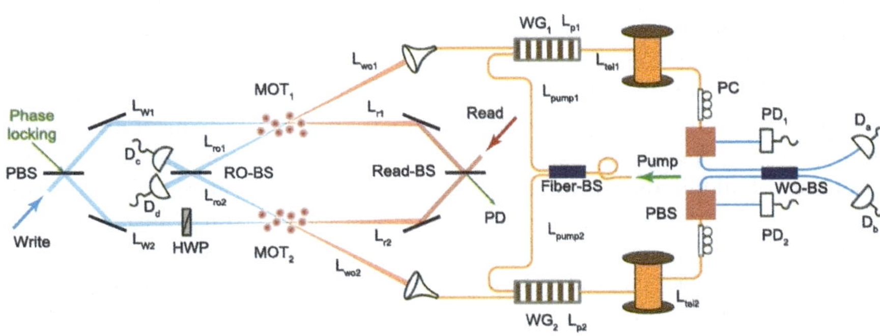

Fig. 7.1 Experimental setup. The single quantum memory still has the ring-cavity enhanced structure

where $|0\rangle_a|0\rangle_p$ represents all atoms locate in the ground states without emitting write-out photons while $|1\rangle_a|1\rangle_p$ indicates that one atom is excited and emitted a write-out photon. p is the excitation probability. When the photons in the entanglement of two such Fock states are frequency converted to 1342 nm by pumping light at 1950 nm in the waveguide, and then interfered with each other at a BS after long distance transmission, the two atomic systems can be projected to the corresponding Fock-type entangled states by single photon detection in the BS output ports. Ignoring higher order events, this process can be written as:

$$\begin{aligned}|\Psi\rangle_{ap1} \otimes |\Psi\rangle_{ap2} &= (|0\rangle_{a1}|0\rangle_{p1} + \sqrt{p}e^{i\varphi_1}|1\rangle_{a1}|1\rangle_{p1})(|0\rangle_{a2}|0\rangle_{p2} + \sqrt{p}e^{i\varphi_2}|1\rangle_{a2}|1\rangle_{p2}) \\ &= |0\rangle_{a1}|0\rangle_{a2}|0\rangle_{p1}|0\rangle_{p2} + \sqrt{p}(e^{i\varphi_1}|1\rangle_{a1}|1\rangle_{p1}|0\rangle_{a2}|0\rangle_{p2} \\ &+ e^{i\varphi_2}|0\rangle_{a1}|0\rangle_{p1}|1\rangle_{a2}|1\rangle_{p2}) + pe^{i(\varphi_1+\varphi_2)}|1\rangle_{a1}|1\rangle_{p1}|1\rangle_{a2}|1\rangle_{p2}.\end{aligned} \quad (7.2)$$

Here φ_1 and φ_2 are the phases accumulated during single-photon transmission. By post-selecting detection events with one and only one write-out photon, the state is rewritten as:

$$|\Psi\rangle_{ap1} \otimes |\Psi\rangle_{ap2} \rightarrow |1\rangle_{a1}|1\rangle_{p1}|0\rangle_{a2}|0\rangle_{p2} + e^{i(\varphi_2-\varphi_1)}|0\rangle_{a1}|0\rangle_{p1}|1\rangle_{a2}|1\rangle_{p2}. \quad (7.3)$$

The atom-atom entangled state corresponding to the detection of a photon after its projection measurement with BS is [6]:

$$|\Psi^{\pm}\rangle = |0\rangle_{a1}|1\rangle_{a2} \pm e^{i\delta\varphi}|1\rangle_{a1}|0\rangle_{a2}, \quad (7.4)$$

where $\delta\varphi = \varphi_2 - \varphi_1 + \pi$. Unlike in two-photon detection, where the prepared entangled state is path independent, the entanglement in the form of Fock states prepared here is correlated with the phase accumulated by the photons on the path, and it also suits the cases when the read process is carried out to verify the correlation between the write-out and read-out photons. In order to obtain a determined entangled state between the write-out and read-out photons, the phase on the path needs to be stabilized, and here we lock the two interferometers to achieve the stable phase.

As shown in Fig. 7.1, the write beam, the read beam, and the pump light for frequency conversion are each prepared from one light through beam splitters. The phases accumulated during the write-out photon detection through the two atomic systems are (the paths with subscripts containing 1 and 2 in the figure):

$$\phi_{wi} = k_w L_{wi} + k_{wo} L_{woi} + k_{tel}(L_{pi} + L_{teli}) - k_p L_{pi}, i = 1, 2. \quad (7.5)$$

The phase in the read process is:

$$\phi_{ri} = k_r L_{ri} + k_{ro} L_{roi}, i = 1, 2. \quad (7.6)$$

The phase introduced here during the light-atom interaction is considered as a fixed value and has been omitted. In order to obtain a definite entangled state, the phase in the above path should satisfy the conditions:

$$\begin{aligned} \phi_{w1} &= \phi_{w2} \\ \phi_{r1} &= \phi_{r2}. \end{aligned} \tag{7.7}$$

During the whole process for entanglement generation and verification, it requires $\phi_{w1} - \phi_{w2} + \phi_{r1} - \phi_{r2} = 0$. In our experiments, we realize this through two separate Mach-Zehnder interferometers, as shown in the figure. An interferometer is composed on the write and read paths, coupling a phase-locked beam into the interferometer through the PBS of write beam. We monitor the interference signal from the other output port of the read BS and feedback to a PZT on the reflector on path $w2$ to stabilize the phase. In subsequent measurements, the phase difference between the experimental and phase-locked light is adjusted on path $w2$ by placing additional QWP, HWP, and QWP to vary the different phases during the verification of entanglement generation. An additional interferometer is composed between the write-out and read-out channels, and the phase-locked light is coupled into the path through one port of the read-out BS using a 1:99 partial reflector with opposite direction. Eventually we also monitor the signal after the BS of write-out interferometer.

7.3 Experimental Results

Experimentally we first carry out experiments without frequency conversion and optimize them, as described in the previous subsection, when the interferometer phase is locked, under the real experimental conditions, the obtained density matrix of states for the read-out photons when a write-out photon is detected after the interference BS is:

$$\rho = c_{00}|00\rangle\langle 00| + c_+|\Psi_p^+\rangle\langle\Psi_p^+| + c_-|\Psi_p^-\rangle\langle\Psi_p^-| + c_{11}|11\rangle\langle 11|. \tag{7.8}$$

Here $\Psi_p^{\pm}$ is for the state with non-zero read-out photons (Eq. 7.4), the zero-photon state comes from irrelevant instances such as dark counting in the write-out detection and non-unity retrieval efficiency during the read-out process, which can be excluded by improving the experiment with a later two-photon detection method. The residual two-photon state comes from a multi-excitation event. $\Psi_p^{\pm}$ corresponds to the state expected under single-photon detection. For measurements under the intrinsic eigen basis, the read-out photons do not interfere, we can defining the visibility in this case as:

$$V_e = \frac{c_+ + c_-}{c_+ + c_- + c_{11}}. \tag{7.9}$$

The measured value in our experiment is 0.969. To verify the entangled state, it is also necessary to measure the result under the superpositional basis by interfering the read-out photons from the two memory systems, and measuring the entanglement visibility V_θ by adjusting the phase θ on the write beam path $w2$. The result is

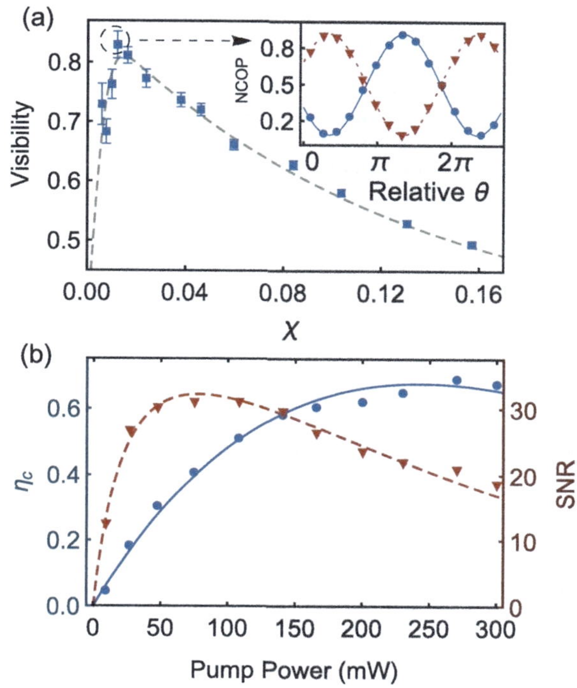

Fig. 7.2 Results for optimizing excitation probability and pumping laser. a the measured entanglement visibility under superpositional basis. Inset shows the normalized probability distribution of the expected photon coincidence counts and its orthogonal correlation. **b** conversion efficiency and SNR with varying power of pumping light

shown in the inset of Fig. 7.2a, where the visibility is calculated as 0.823. Finally, the entanglement fidelity after post-selecting the non-zero detection events can be calculated as:

$$F = \frac{1}{2}(V_e + V_\theta) = 0.896. \tag{7.10}$$

Going further, we measure the effect of the excitation probability on the visibility of the entanglement under the superpositional basis, as shown in Fig. 7.2a. If the excitation probability is too high, multi-excitation events will lead to the ratio increase of two-photon detection, if the excitation probability is too low, the influence of dark counting events and noise photons become more obvious. Both cases will affect the entanglement quality, after making a tradeoff, we set the excitation probability to 0.02. At this point, the quality of the entanglement is mainly limited by the fidelity of the photon interference. We perform HOM interference on the write-out photons and read-out photons from the two quantum memories, respectively. The measured interference visibility is about 14:1 and 11:1.

After the local measurements were completed, we started to add the frequency conversion part of the device and measured the conversion efficiency and signal-to-noise ratio with the pumping light power as shown in Fig. 7.2b, where the conversion

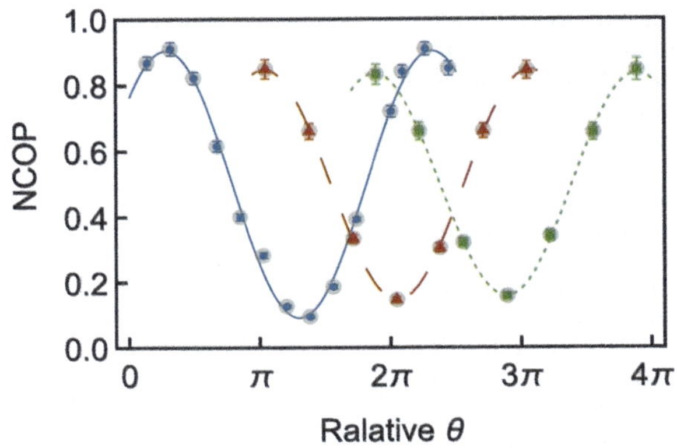

Fig. 7.3 The variation curve of the probability of coincidences of the expected correlated photons measured under the superpositional basis at different distances with the varying phase. Blue 10 m. Red: 10 km. Green: 50 km

Table 7.1 Results for the case of whether performing quantum frequency conversion and the case of different distances with quantum frequency conversion

	c_{00}	$c_+ + c_-$	c_{11} (‰)	V_θ	F
no QFC(5 m)	0.679(9)	0.311(7)	9.95(42)	0.823(19)	0.896(18)
QFC(10 m)	0.692(9)	0.304(7)	3.92(49)	0.812(19)	0.900(21)
QFC(10 km)	0.714(6)	0.282(4)	4.36(30)	0.699(8)	0.842(13)
QFC(50 km)	0.725(8)	0.271(6)	3.87(42)	0.677(9)	0.832(19)

efficiency is low when the power is too low and the noise become higher when the power is too high. Experimentally the noise mainly comes from the Raman noise of the pumping light inside the waveguide, the spectrum range is in the 1342 ± 2.5 nm range, which can be filtered by a band-pass filter, the noise generated by frequency doubling and tripling is also able to be filtered by the corresponding filter, and the residual noises can then be filtered by the WDM. As a result, we are able to obtain a conversion efficiency of about 0.33 (including the internal conversion efficiency) 0.7, filtering efficiency 0.8 and fiber coupling efficiency 0.6) with a signal-to-noise ratio of about 20:1.

After combining the quantum frequency conversion, we test the entanglement visibility under the superpositional basis at three different distances as shown in Fig. 7.3 and Table 7.1. The fidelity of the entangled state we obtain keeps the same with or without the addition of the frequency conversion device, indicating that the whole frequency conversion process preserves the photonic quantum properties and does not affect the quality of the final obtained entangled state. The write-out photon loss in the fiber after quantum frequency conversion is reduced to 0.3 dB/km, and when the distance is increased to 10 and 50 km, the photon counting rate is about 1 counts/s, 0.14 counts/s, and the corresponding fidelity is 0.842 (13) and 0.832 (19), which is sufficient to prove that we have successfully obtained entanglement between two quantum memories at a long distance experimentally.

7.4 Summary

In this chapter, we utilize the efficient cold atomic quantum memory and combine quantum frequency conversion technology to reduce the single-photon transmission loss to 0.3 dB/km, thus making entanglement of remote quantum nodes feasible. The detection of a single photon can herald the establishment of entanglement between cold atomic quantum memories with a heralding efficiency of nearly 100%, and finally we successfully realize the entanglement between two quantum memories separated by fiber transmission over dozens of kilometers. It is the longest fiber length for entangled quantum memories achieved in the world, our work lays the foundation for the future construction of global quantum networks.

References

1. Radnaev AG, Dudin YO, Zhao R, Jen HH, Jenkins SD, Kuzmich A, Kennedy TAB (2010) A quantum memory with telecom-wavelength conversion. Nat Phys 6(11):894
2. Dudin YO, Radnaev AG, Zhao R, Blumoff JZ, Kennedy TAB, Kuzmich A (2010) Entanglement of light-shift compensated atomic spin waves with telecom light. Phys Rev Lett 105(26):260502
3. Albrecht Boris, Farrera Pau, Fernandez-Gonzalvo Xavier, Cristiani Matteo, De Riedmatten Hugues (2014) A waveguide frequency converter connecting rubidium-based quantum memories to the telecom c-band. Nat Commun 5:3376
4. Maring Nicolas, Farrera Pau, Kutluer Kutlu, Mazzera Margherita, Heinze Georg, de Riedmatten Hugues (2017) Photonic quantum state transfer between a cold atomic gas and a crystal. Nature 551(7681):485
5. Chou C-W, De Riedmatten H, Felinto D, Polyakov SV, Van Enk SJ, Jeff Kimble H (2005) Measurement-induced entanglement for excitation stored in remote atomic ensembles. Nature 438(7069):828
6. Duan L-M, Lukin MD, Ignacio Cirac J, Zoller P (2001) Long-distance quantum communication with atomic ensembles and linear optics. Nature 414(6862):413

Chapter 8
Summary and Outlook

This thesis is a summary of my research work during the last few years of my Ph.D. The main system of research is quantum memory based on cold atomic ensembles. Flying photons can be used to transmit quantum information over long distances, and quantum memories based on cold atomic ensembles can act as carriers for local storage and manipulation of quantum information, so it is of great significance to study cold atomic quantum memories for construction of global quantum networks. In this thesis, based on previous work, we further scrutinize the factors that will affect the performance metrics of quantum memory such as retrieval efficiency, and then we combine these analysis with some innovative experimental techniques, which will enable us to finally realize a highly retrievable quantum memory with intrinsic retrieval efficiency up to 90% and overall measured retrieval efficiency up to 40%. Based on this efficient quantum memory, we optimize the experimental loading time for preparing cold atomic ensembles, the number of experimental trials in each loading and other timing parameters, so that the experimental detection brightness of photons can support our experimental study of multiple quantum memories.

Going further, we move from a single photonic quantum memory to realization of entanglement between the polarized state of photons and different Zeeman splitting spin states of atoms, and analyze in detail the influencing effects on the fidelity of the entanglement in terms of the photonic polarizations, magnetic fields, pumping processes, and multiple excitation events. In the end, we are able to achieve an entangled photon-atom state with a fidelity of 0.926 after optimizing these influencing parameters. The remaining limiting factors is also discussed for better understanding of our experimental system. After successfully generating such a high performance source of photon-atom entanglement, we start to realize entanglement between two cold atomic quantum memories by interfering the two write-out photons to achieve entanglement swapping, where the fidelity of the entanglement swapping process is strictly related to the indistinguishability of the interfered photons. So we experimentally analyze in detail the influencing factors of indistinguishability of the write-out photons and combine some innovative settings to obtain an entangled state with

B. Jing, *Quantum Network with Multiple Cold Atomic Ensembles*, Springer Theses,
https://doi.org/10.1007/978-981-19-0328-1_8

a fidelity of 0.85. After a thorough study of the photon interference problem, we are able to connect multiple quantum memories to achieve one of the simplest two-dimensional quantum networks. Both the entanglement of three separate cold atomic quantum memories and the hybrid six-fold entanglement of three cold atomic quantum memories and three flying write-out photons have been reported for the first time. The generated states are all maximum entangled GHZ states, which are of great significance for building global quantum networks. In order to extend the distance between adjacent network nodes, we combine quantum frequency conversion technology to convert photons generated by the atoms in the near-infrared band into photons in the telecom band with much lower fiber transmission loss, thus allowing us to achieve entanglement between two quantum memories at a fiber distance of 50 km, which is also the longest distance achieved in the world so far.

In addition, with the help of the quantum memories, we have simultaneously generated three photons with fully distinguishable frequencies, the generation process is superior to the current main resort to the parametric down-conversion process, both in terms of the scalability of the simultaneously generated photons and the difference in frequency. Using the generated photons, we have achieved for the first time the interference of multiple frequency distinguished photons in a network, which lays the foundation for connecting quantum networks composed of different physical systems and also reduces the requirement of having to use identical photons in conventional boson sampling. Our work provides a research direction for future quantum information processing based on distinguished photons.

In the future experiments, our biggest goal is to be able to construct a global quantum network. In current experiments, in order to guarantee the quality of the photon-atom entanglement source, we have to control the excitation probability at a very low level, which makes each generation process for photon-atom entanglement not efficient enough, as a solution, photon-atom entanglement based on the Rydberg blockade mechanism does not have to worry about this problem, the whole process will become deterministic, and both the efficiency of the entanglement generation and the quality of the entanglement will be better than now. Considering the distance extension among remote quantum nodes, strict demands are made on the storage lifetime of the quantum memory, one possible way is to restrict the atomic motions by trapping the atoms in a three-dimensional optical lattice, thus enabling storage lifetime up to sub-seconds. Although atom-atom entanglement based on single photon detection has been achieved right now, as the number of network nodes increases, the effect of the vacuum state component in the Fock-type entanglement becomes increasingly significant and will make the entanglement limited use for practical applications in quantum repeater and further quantum networking, so eventually we will need to find efficient ways for supporting us to carry out entangled connections in a two-photon detection configuration. And for photon transmission loss over long fibers, because of the low fiber transmission photon loss in the telecom band, the correlation of atomic quantum memories with photons in the telecom band has to be obtained. Nowadays it is usually obtained by means of quantum frequency conversion, and the total efficiency of the conversion process becomes a new and demanding indicator. Given that the rubidium atoms we use contain energy level transitions in the telecom band within themselves, directly generating entanglement between atomic quantum memories and telecom band photons with the help of these transitions is also a very interesting direction of research.

Appendix A
Saturated Absorption Spectrum for ^{87}Rb D1 and D2 Line

The figures are cited from the Sacher Lasertrchnik group.

B. Jing, *Quantum Network with Multiple Cold Atomic Ensembles*, Springer Theses,
https://doi.org/10.1007/978-981-19-0328-1

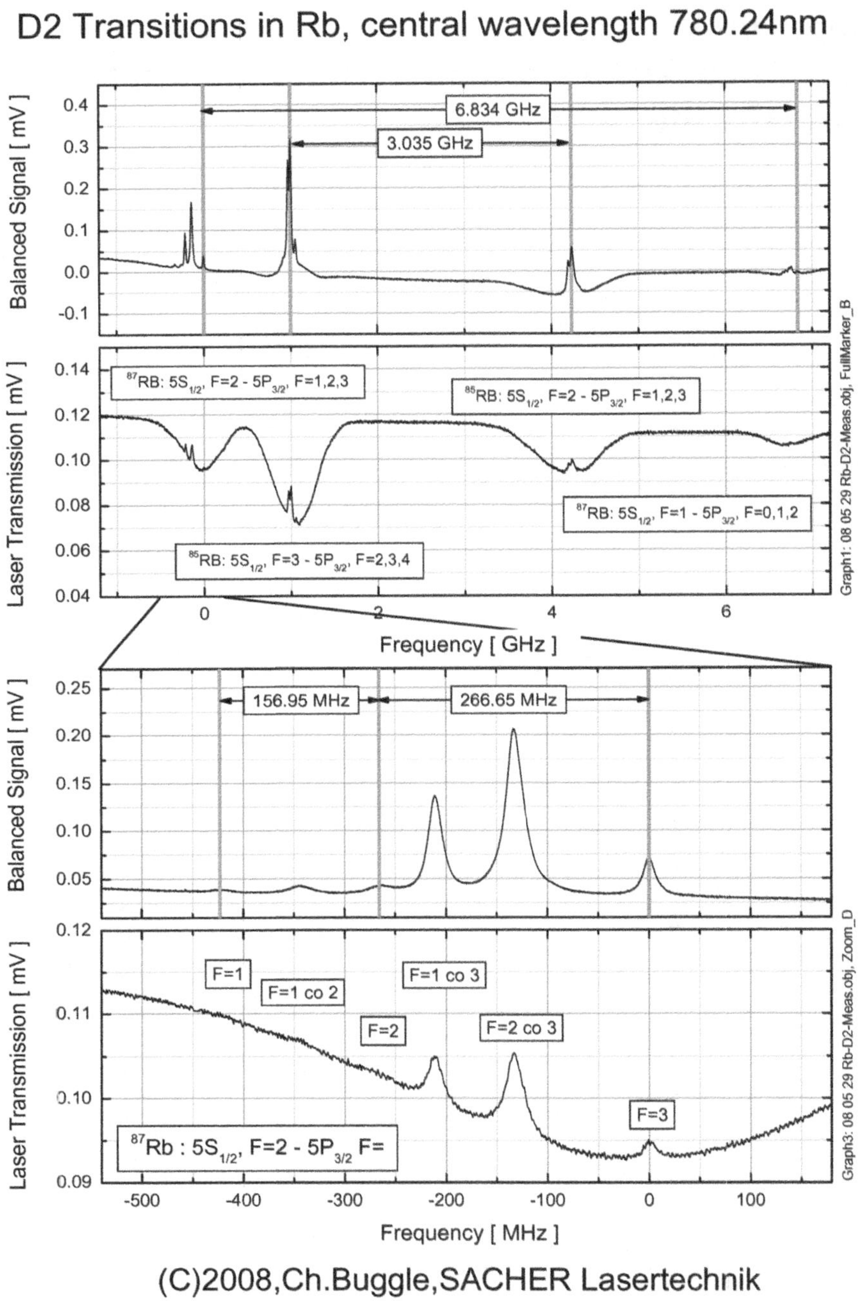
D2 Transitions in Rb, central wavelength 780.24nm
Balanced Signal [mV]
6.834 GHz
3.035 GHz
Laser Transmission [mV]
87RB: 5S1/2, F=2 - 5P3/2, F=1,2,3
85RB: 5S1/2, F=2 - 5P3/2, F=1,2,3
87RB: 5S1/2, F=1 - 5P3/2, F=0,1,2
85RB: 5S1/2, F=3 - 5P3/2, F=2,3,4
Frequency [GHz]
Graph1: 08 05 29 Rb-D2-Meas.obj, FullMarker_B
Balanced Signal [mV]
156.95 MHz
266.65 MHz
Laser Transmission [mV]
F=1
F=1 co 2
F=1 co 3
F=2
F=2 co 3
F=3
87Rb : 5S1/2, F=2 - 5P3/2 F=
Frequency [MHz]
Graph3: 08 05 29 Rb-D2-Meas.obj, Zoom_D
(C)2008,Ch.Buggle,SACHER Lasertechnik

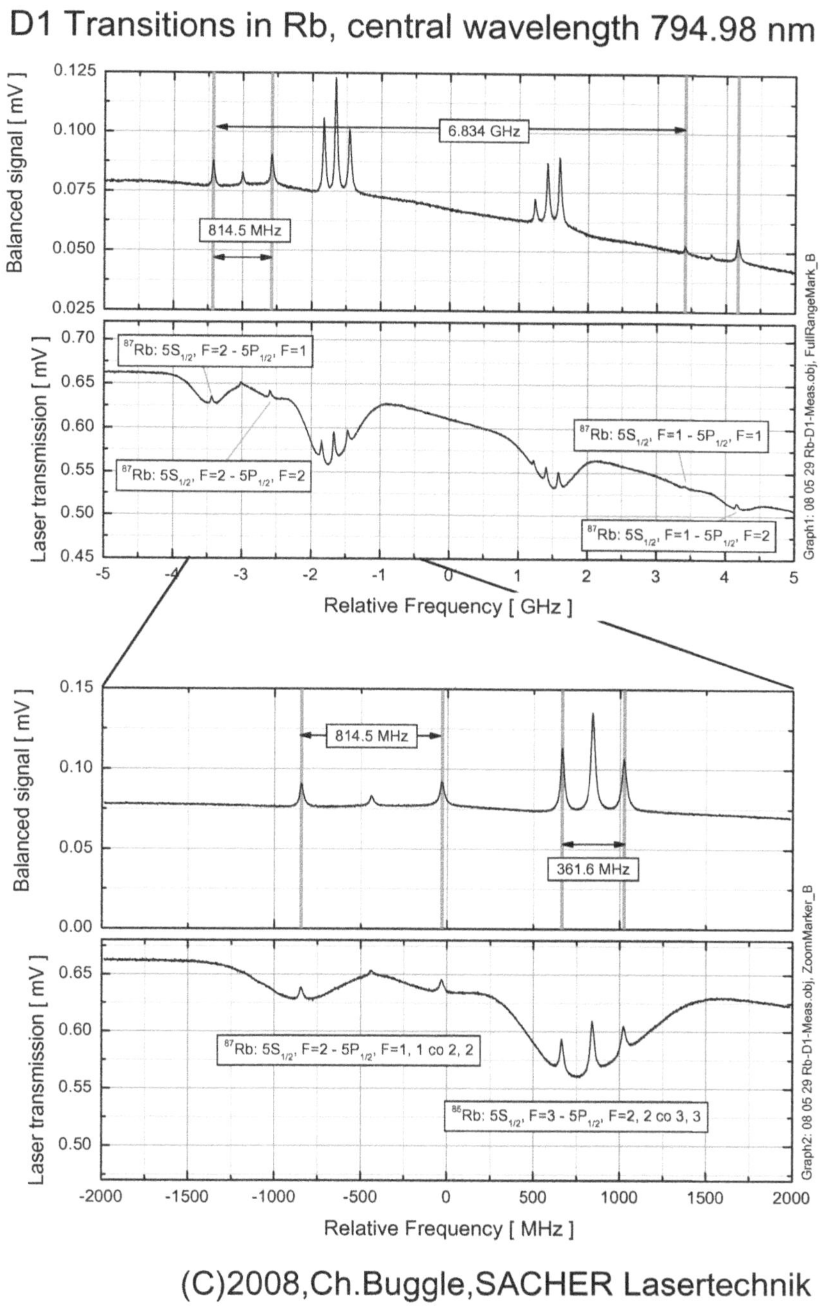
D1 Transitions in Rb, central wavelength 794.98 nm
Balanced signal [mV]
Laser transmission [mV]
6.834 GHz
814.5 MHz
87Rb: 5S1/2, F=2 - 5P1/2, F=1
87Rb: 5S1/2, F=2 - 5P1/2, F=2
87Rb: 5S1/2, F=1 - 5P1/2, F=1
87Rb: 5S1/2, F=1 - 5P1/2, F=2
Relative Frequency [GHz]
Graph1: 08 05 29 Rb-D1-Meas.obj, FullRangeMark_B
814.5 MHz
361.6 MHz
87Rb: 5S1/2, F=2 - 5P1/2, F=1, 1 co 2, 2
85Rb: 5S1/2, F=3 - 5P1/2, F=2, 2 co 3, 3
Relative Frequency [MHz]
Graph2: 08 05 29 Rb-D1-Meas.obj, ZoomMarker_B
(C)2008,Ch.Buggle,SACHER Lasertechnik

Appendix B
Calculation of Magnetic Field Generated by Rectangular Helmholtz Coils

Here we consider two rectangular coils (Fig. B.1) with side lengths of $2a$ and $2b$ and a spacing of $2H$. The joint center of the two coils is set to the origin point. After applying a current of I, we calculate the intensity of the magnetic field generated along the direction of z at the point (x, y, z).

According to the Biot–Savart law, the magnetic fields generated by the four sides of the first coil in the z direction are as follows:

$$B_1(z) = \frac{\mu_0 I}{4\pi}\frac{b-x}{(b-x)^2+(z+H)^2}\left[\frac{a+y}{\sqrt{(a+y)^2+(b-x)^2+(z+H)^2}}+\frac{a-y}{\sqrt{(a-y)^2+(b-x)^2+(z+H)^2}}\right], \tag{B.1}$$

$$B_2(z) = \frac{\mu_0 I}{4\pi}\frac{a-y}{(a-y)^2+(z+H)^2}\left[\frac{b-x}{\sqrt{(a-y)^2+(b-x)^2+(z+H)^2}}+\frac{b+x}{\sqrt{(a-y)^2+(b+x)^2+(z+H)^2}}\right], \tag{B.2}$$

$$B_3(z) = \frac{\mu_0 I}{4\pi}\frac{b+x}{(b+x)^2+(z+H)^2}\left[\frac{a+y}{\sqrt{(a+y)^2+(b+x)^2+(z+H)^2}}+\frac{a-y}{\sqrt{(a-y)^2+(b+x)^2+(z+H)^2}}\right], \tag{B.3}$$

$$B_4(z) = \frac{\mu_0 I}{4\pi}\frac{a+y}{(a+y)^2+(z+H)^2}\left[\frac{b-x}{\sqrt{(a+y)^2+(b-x)^2+(z+H)^2}}+\frac{b+x}{\sqrt{(a+y)^2+(b+x)^2+(z+H)^2}}\right]. \tag{B.4}$$

The sum of the magnetic field intensity generated by the coil in the z direction is:

$$B_d(z) = B_1(z) + B_2(z) + B_3(z) + B_4(z). \tag{B.5}$$

Similarly, the magnetic field generated by the other coil in the z direction is as follows:

$$B_5(z) = \frac{\mu_0 I}{4\pi}\frac{b-x}{(b-x)^2+(z-H)^2}\left[\frac{a+y}{\sqrt{(a+y)^2+(b-x)^2+(z-H)^2}}+\frac{a-y}{\sqrt{(a-y)^2+(b-x)^2+(z-H)^2}}\right], \tag{B.6}$$

$$B_6(z) = \frac{\mu_0 I}{4\pi}\frac{a-y}{(a-y)^2+(z-H)^2}\left[\frac{b-x}{\sqrt{(a-y)^2+(b-x)^2+(z-H)^2}}+\frac{b+x}{\sqrt{(a-y)^2+(b+x)^2+(z-H)^2}}\right], \tag{B.7}$$

B. Jing, *Quantum Network with Multiple Cold Atomic Ensembles*, Springer Theses,
https://doi.org/10.1007/978-981-19-0328-1

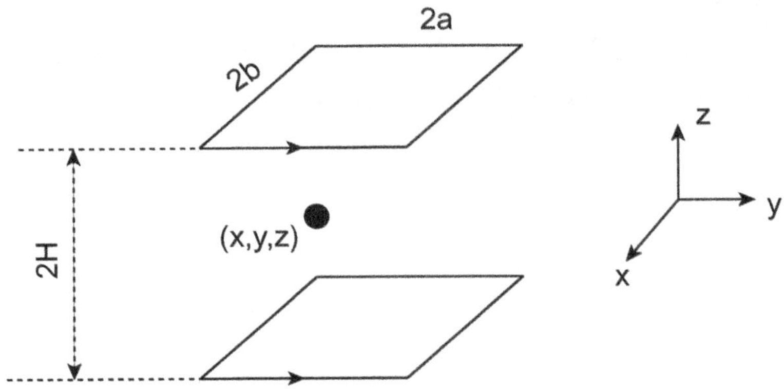

Fig. B.1 Schematic diagram of the magnetic field generated by a pair of rectangular coils

$$B_7(z) = \frac{\mu_0 I}{4\pi}\frac{b+x}{(b+x)^2+(z-H)^2}\left[\frac{a+y}{\sqrt{(a+y)^2+(b+x)^2+(z-H)^2}}+\frac{a-y}{\sqrt{(a-y)^2+(b+x)^2+(z-H)^2}}\right], \tag{B.8}$$

$$B_8(z) = \frac{\mu_0 I}{4\pi}\frac{a+y}{(a+y)^2+(z-H)^2}\left[\frac{b-x}{\sqrt{(a+y)^2+(b-x)^2+(z-H)^2}}+\frac{b+x}{\sqrt{(a+y)^2+(b+x)^2+(z-H)^2}}\right]. \tag{B.9}$$

$$B_u(z) = B_5(z) + B_6(z) + B_7(z) + B_8(z). \tag{B.10}$$

When the currents in the two coils are around the same direction, the Helmholtz coil is mainly used to generate a uniform magnetic field. The direction of the magnetic field is the same at the origin ($x = y = 0$).

$$B(z) = B_d(z) + B_u(z), \tag{B.11}$$

with $\partial B(z)/\partial(z)|_{z=0} = 0$. For square coils, $a = b = l$,

$$\begin{aligned} B(z) &= B_z(x=0, y=0, z) \\ &= \frac{2\mu_0 I l^2}{\pi[l^2+(z+H)^2]\sqrt{2l^2+(z+H)^2}} + \frac{2\mu_0 I l^2}{\pi[l^2+(z-H)^2]\sqrt{2l^2+(z-H)^2}}. \end{aligned} \tag{B.12}$$

When the directions of the current in the coils are opposite, they are the so-called anti-Helmholtz coils, a gradient magnetic field can be generated at the origin, such as the magnetic field in a magneto-optical trap, the difference is that a pair of circular coils is used in our experiment.

Appendix C
Retrieval Efficiency Under Different Polarization Combinations

Note: Even though the efficiency here is tested in the very beginning of our experiment and is not the best performance we can achieve later, the relative values can still give the compared information of different settings, i.e., the retrieval efficiency varies with the polarization combinations (Tables C.1 and C.2).

Table C.1 The relative comparison of retrieval efficiency under different polarization combinations of write beam, write-out photon, read beam, and read-out photon when the initial state of the atoms is prepared in the ground state $F = 1$

Polarization combination	Retrieval efficiency	Polarization combination	Retrieval efficiency
RRRR	0.213	LRRL	0.195
RRLL	0.194	HVVH	0.185
HVHV	0.110	VHHV	0.179
VHVH	0.082	–	–

Table C.2 The relative comparison of retrieval efficiency under different polarization combinations of write beam, write-out photon, read beam, and read-out photon when the initial state of the atoms is prepared in the ground state $F = 1$ and whether π-pump laser is applied to prepare atoms into $F = 1, m_F = 0$ (The efficiency is roughly optimized compared to that in Table C.1)

Polarization combination	Retrieval efficiency	Polarization combination	Retrieval efficiency
HVVH	0.235	–	–
VHHV	0.255	VHHV(Pi)	0.237
LLLL	0.247	LLLL(Pi)	0.224
LRRL	0.163	LRRL(Pi)	0.207

B. Jing, *Quantum Network with Multiple Cold Atomic Ensembles*, Springer Theses,
https://doi.org/10.1007/978-981-19-0328-1

Appendix D
Fidelity Estimation of Bell State and GHZ State

Bell state.[1] In our experiment, the atom-photon entangled state is written as

$$|\Psi\rangle_{ph-a} = \frac{1}{\sqrt{2}} \left(|\sigma^+ \downarrow\rangle + e^{i\varphi(t)} |\sigma^- \uparrow\rangle \right). \tag{D.1}$$

At a specific storage time $\varphi(t) = 0$, we retrieve the atomic state to a single read-out photon, then the write- and read-out photon pairs become to a familiar form

$$|\Psi^+\rangle = \frac{1}{\sqrt{2}} \left(|\sigma^+ \sigma^-\rangle + |\sigma^- \sigma^+\rangle \right). \tag{D.2}$$

When we project the experimental state ρ_{exp} to the ideal one ρ_{ide}, the state fidelity can be expressed as a function of entanglement visibilities,

$$F(\rho_{ide}, \rho_{exp}) = \mathrm{Tr}\left[|\Psi^+\rangle\langle\Psi^+| \rho_{exp} \right] = \frac{1}{4} \left(1 + \mathrm{V_{RL}} + \mathrm{V_{HV}} + \mathrm{V_{DA}} \right), \tag{D.3}$$

where $\mathrm{V_{RL}}$ is the eigen bases $\sigma^\pm$ visibility, and $\mathrm{V_{HV}}$ ($\mathrm{V_{DA}}$) is the superpositional bases $\sigma^+ \pm \sigma^-$ ($\sigma^+ \pm i\sigma^-$) visibility. According to our experimental results, $\mathrm{V_{HV}} \simeq \mathrm{V_{DA}}$, therefore the fidelity $F(\rho_{ide}, \rho_{exp})$ is estimated as $(1 + \mathrm{V_{RL}} + 2\mathrm{V_{HV}})/4 = 0.926$.

GHZ state. Usually, to prove multipartite entangled state, the entanglement witness method is used.[2] Consider an observable $\mathcal{W}$, for all separable states ρ_s, the expectation value of $\mathcal{W}$ is nonnegative, if there exits an entangled state ρ_e, whose expectation value is negative, then $\mathcal{W}$ is called an entanglement witness. For the $|\mathrm{GHZ}_6\rangle$ state, the witness $\mathcal{W}$ is defined as follows,

[1] Also see supplementary information of Nature Photonics, 13, 210–213 (2019).

[2] Physical Review Letters 92, 087902 (2004). Physics Reports 474, 1–75 (2009).

B. Jing, *Quantum Network with Multiple Cold Atomic Ensembles*, Springer Theses,
https://doi.org/10.1007/978-981-19-0328-1

$$\mathcal{W}_6 = \frac{I}{2} - |\mathrm{GHZ}_6\rangle\langle\mathrm{GHZ}_6|, \tag{D.4}$$

where I is the 64×64 identity operator. Once the measured expectation value of $\mathcal{W}_6$ is negative, the existence of entangled state $|\mathrm{GHZ}_6\rangle$ can be proved. The density operator $|\mathrm{GHZ}_6\rangle\langle\mathrm{GHZ}_6|$ can be decomposed into locally measurable observables,

$$\begin{aligned}&|\mathrm{GHZ}_6\rangle\langle\mathrm{GHZ}_6| = \\ &\frac{1}{2}(|HHH\downarrow\downarrow\uparrow\rangle\langle HHH\downarrow\downarrow\uparrow| + |VVV\uparrow\uparrow\downarrow\rangle\langle VVV\uparrow\uparrow\downarrow|) + \frac{1}{12}\sum_{n=0}^{5}(-1)^n M_n^{\otimes 6},\end{aligned} \tag{D.5}$$

where $M_n = \cos(n\pi/6)\sigma_x + \sin(n\pi/6)\sigma_y$. For photons, Pauli matrices are defined as:

$$\sigma_x = |D\rangle\langle D| - |A\rangle\langle A|, \quad \sigma_y = |R\rangle\langle R| - |L\rangle\langle L|, \quad \sigma_z = |H\rangle\langle H| - |V\rangle\langle V|, \tag{D.6}$$

where $|D/A\rangle = (|H\rangle \pm |V\rangle)/\sqrt{2}$, $|R/L\rangle = (|H\rangle \pm i|V\rangle)/\sqrt{2}$. Similarly, for the third atomic qubit, Pauli matrices are defined as:

$$\sigma_x = |\rightarrow\rangle\langle\rightarrow| - |\leftarrow\rangle\langle\leftarrow|, \quad \sigma_y = |\circlearrowright\rangle\langle\circlearrowright| - |\circlearrowleft\rangle\langle\circlearrowleft|, \quad \sigma_z = |\uparrow\rangle\langle\uparrow| - |\downarrow\rangle\langle\downarrow|, \tag{D.7}$$

where $|\rightarrow / \leftarrow\rangle = (|\uparrow\rangle \pm |\downarrow\rangle)/\sqrt{2}$, $|\circlearrowright / \circlearrowleft\rangle = (|\uparrow\rangle \pm i|\downarrow\rangle)/\sqrt{2}$. While for the first two atomic qubits, the $|\downarrow\rangle/|\uparrow\rangle$ in the definition is flipped. The expectation value of witness is calculated as,

$$\begin{aligned}\mathrm{Tr}\left[\mathcal{W}\rho_{exp}\right] &= \frac{1}{2} - \frac{1}{2}\mathrm{Tr}\left[(|HHH\downarrow\downarrow\uparrow\rangle\langle HHH\downarrow\downarrow\uparrow| + |VVV\uparrow\uparrow\downarrow\rangle\langle VVV\uparrow\uparrow\downarrow|)\rho_{exp}\right] \\ &\quad - \frac{1}{12}\sum_{n=0}^{5}(-1)^n\mathrm{Tr}\left[M_n^{\otimes 6}\rho_{exp}\right] \\ &= \frac{1}{2} - \frac{1}{2}\langle M_{HV}\rangle - \frac{1}{12}\sum_{n=0}^{5}(-1)^n\langle M_n^{\otimes 6}\rangle,\end{aligned} \tag{D.8}$$

where $\langle M_{HV}\rangle$ is the measured normalized six-fold coincidence probability of $|HHH\downarrow\downarrow\uparrow\rangle$ or $|VVV\uparrow\uparrow\downarrow\rangle$ (Fig. 5.17a) at eigen bases, $\langle M_n^{\otimes 6}\rangle$ is the measured visibility (Fig. 5.17b) at different superpositional bases ($|H\rangle \pm e^{in\pi/6}|V\rangle$ for photon qubits, $|\downarrow\rangle \pm e^{in\pi/6}|\uparrow\rangle$ or $|\uparrow\rangle \pm e^{in\pi/6}|\downarrow\rangle$ for atomic qubits). The measured expectation value of witness is -0.186, which proves that the genuine six-party entanglement is generated successfully. From the measured witness $\mathcal{W}$, the fidelity of the experimental state can be derived directly $F_{\mathrm{GHZ}_6} = \langle\mathrm{GHZ}_6|\rho_{exp}|\mathrm{GHZ}_6\rangle = 1/2 - \mathrm{Tr}\left[\mathcal{W}\rho_{exp}\right] = 0.686$.

The manufacturer's authorised representative in the EU is Springer Nature Customer Service Centre GmbH, Europaplatz 3, 69115 Heidelberg, Germany. If you have any concerns regarding our products, please contact ProductSafety@springernature.com

Printed and bound by CPI Group (UK) Ltd, Croydon, CR0 4YY
30/09/2026
02225245-0002